Editor
Kim Fields

Editorial Project Manager
Mara Ellen Guckian

Managing Editors
Karen J. Goldfluss, M.S. Ed.
Ina Massler Levin, M.A.

Illustrators
Blanca Apodaca
Janice Kubo
Sue Fullam

Cover Artist
Barb Lorseyedi

Art Production Manager
Kevin Barnes

Imaging
Craig Gunnell
James Edward Grace

Publisher
Mary D. Smith, M.S. Ed.

Written by

Lynne R. Weaver, M.S.

Teacher Created Resources, Inc.
6421 Industry Way
Westminster, CA 92683
www.teachercreated.com

ISBN-1-4206-3184-5

©2006 Teacher Created Resources, Inc.

Made in U.S.A.

Table of Contents

Introduction

Kindergarten has changed a great deal in the last 20 years. Many baby boomers remember playing dress up, building with blocks, and singing the ABC's. These activities are still part of kindergarten rooms, but today the classroom routine has become much more academic. Children are expected to have an understanding of numbers, patterns, money, and the concept of time by the end of their first formal year of school.

Preschool teachers have had to adjust their curriculums to ensure that they are adequately preparing their children for kindergarten. Many districts now have published standards and benchmarks that preschool teachers can use as a reference for the math skills that need to be taught. Still, having a new list of standards can be overwhelming. Where do you start?

Year Round Preschool Math defines 26 math objectives for young children. This book is filled with games, crafts, home/school activities, literature suggestions, and manipulative ideas that are new and exciting to preschoolers. Each suggested activity relates to the featured educational objective. Ideas presented in one section can be adapted or expanded to use with other math activities.

Math is part of everyday life. The activities included in this book will help to enhance life experiences or draw attention to them in new ways. It is also important to note that math experiences do not always have to be planned to be meaningful. As teachers, we try to be alert to the "teaching moment" that may present itself during the day. For instance, when a child returns from a trip to the beach and brings a shell for each classmate, it is a great time to discuss one-to-one correspondence. Has the child brought enough shells for everyone to have one? How will he or she know? What about a snack that can be divided in half by the child before it is eaten? Try to use events like these to reinforce the math skills you have already discussed or to introduce the children to something new.

Preschool education should be a partnership between the school and the home. Let parents know the math skills being introduced in the classroom. This can be done using a class newsletter, daily report, or information posted on a bulletin board. Occasionally, send home a fun activity to be completed with the assistance of the family. (Suggested parent letters are included in this book.)

Children participating in this curriculum will have fun learning new math skills. These future leaders will be approaching math with a new level of confidence and will be better prepared for the next level of education.

The Five L's for Math

It is helpful to employ the five L's when teaching specific math concepts to preschoolers.

Layout

Layout refers to the setup of the classroom. The learning environment should be inviting, math-friendly, and exciting for the children. A Math Center is an integral part of the math curriculum and should be available to the children every day. The materials in this area should include manipulatives and games that will reinforce what is being taught in the classroom. Posters and other visuals can also be displayed to enhance the Math Center. The teacher is responsible for introducing new items and removing items for a time and then returning them. A description of a typical Math Center in the classroom is described on page 6.

Language

Math language should be modeled by the teacher so the children begin to incorporate it into their own daily vocabulary. Always use the appropriate terms associated with each skill being taught. Refer to the Vocabulary section at the beginning of each unit for sample terms. For example, when asking a child to sort some items, the teacher might say, "Can you sort these clocks into two groups? Can you classify them?" When a child reports spontaneously to a teacher, "Look, I sorted the paper clips by colors!", he or she has incorporated math language into everyday language.

Listening Activities

Listening activities are another great way for children to practice recently learned skills. These activities should be a part of everyday routine. The repetition of ideas shared during these activities will help reinforce the skills for children at different levels. Listening Activities are described in more detail on page 7. These activities can also be used to evaluate if each child is grasping the concepts being taught. The advantage of this teaching technique is that children continue to learn as they watch and listen to their peers' responses. Use the Listening Activities included in each unit to give children another chance to learn or review math concepts.

Literature

Math literature can be used to introduce children to many new math concepts. In addition, literature can be used in a variety of ways to supplement the teaching goals. A separate section titled "How to Use the Literature Selections" can be found on page 8. Refer to the Selected Literature list in each unit for books that will reinforce concepts learned in that unit.

Laughter

Laughter is the last, but not the least, important L. Laugh and have fun with math. Math doesn't have to be a serious undertaking; it can be lighthearted and enjoyable! Make laughter a natural part of every math lesson. The more enthusiasm you show, the more the children will have positive experiences with math.

How to Use This Book

The information in *Year Round Preschool Math* is to be used as a reference guide to help inspire teachers to provide an enriching math program for their preschoolers. A chart spreading the goals out over a suggested nine-month period can be found on page 9. This chart can be used throughout the year as a quick reference guide. Teachers can adjust the goals to meet the specific needs of a particular class. The key is to proceed slowly and to advance as the children show evidence of understanding the concepts.

The next few pages are used to describe the five L's in greater detail. The layout of the classroom, specifically a Math Center area, is described on page 6. This is followed by a more in-depth description of the Listening Activities. The next section describes how to incorporate the literature selections that are provided for each goal. The remainder of the book presents 26 math objectives and a list of activities to reinforce each one. Each unit includes the following components:

Educational Objectives—The educational objectives provided describe what the child should be able to do upon entering kindergarten. The activities provided in the units should help children develop the skills to meet these needs.

Vocabulary—The math language, or vocabulary, needed for each objective is detailed here. Incorporate these terms into daily language use whenever possible. Encourage children to do likewise. Review the vocabulary words on a regular basis.

Classroom Activities—Circle ideas, art projects, and games are described here. Directions for assembling suggested games are supplied and patterns are provided on subsequent pages in each unit. Once assembled, these unit-specific games can be added to the Games bin in the Math Center. The Math Center and circle time are excellent places to practice math.

Math skills can also be reinforced during snack time. Suggestions may be given to help the child learn the math concepts using the food offered as a snack.

Listening Activities—Specific one-on-one activities are described that can be presented during large group times. Children listen closely to the question or the presentation of the activity and then take turns responding. Children they listen to each peer's response. In the beginning, children will need help taking turns. Some may be shy and need encouragement. Soon though, most will look forward to these daily activities and grow comfortable listening and sharing information.

Selected Literature—A list of recommended books is included for each unit. These books can be used in a variety of ways to help children grasp the math concepts that are presented to them. At the back of the book (pages 235–236) is a section titled, "Oldies, but Goodies." If you can find them, these books will be wonderful additions to your math library. The books in this section are not currently in print, but can be found at many local libraries or used bookstores.

The Math Center

Establishing a good layout for a Math Center is not as difficult as it sounds. The goal is to create an inviting space where children can enjoy using different materials to increase their math knowledge.

Here are some suggestions to help establish a layout for a Math Center.

- Arrange a table with seating where four children can work/play.
- Establish a shelf area to display games and manipulatives. Place word and picture labels on the shelves to mark where each item belongs when not in use. Catalog pictures work well for this type of labeling. Make certain the materials are arranged neatly and are not crowded.
- Label a special basket or bin for "homemade" games. Most of the games suggested in this book can be kept in plastic pouches or one-gallon plastic, resealable bags. Rotate the games to correlate with the unit of focus.
- Display a large, laminated number line with numerals 1–30. (See pages 237–240 for a number line.) If possible, attach the number line to a long table.
- Post a large, laminated poster with the numerals 1–100. Place the poster in the center so that the children can touch the numbers. Allow space for additional posters depicting assorted math concepts, to be added to the Math Center throughout the year.
- Make space for a bulletin board, magnetic chalkboard, and/or flannelboard.
- Find a storage area for materials not in use. Materials can be stored in labeled baskets or boxes.

The Math Center should have only two or three items set out during the first week of use. During whole group time, introduce the children to the center and demonstrate the proper way to care for the materials. Discuss appropriate behavior in the center. Establish a set of rules for participation and care of the center. Explain that new items will be introduced gradually and that different materials will be available for each new unit.

Use the Math Center to reinforce new skills, play games in small groups, encourage open-ended exploration, and practice with math manipulatives. Model how to play each game and use each manipulative before it is added to the center. Include new math language during each demonstration. Younger children may need to play with the assistance of a teacher. It is also a good idea to limit the number of games put out at any one time. Rotate the games so that previous skills are reviewed on a regular basis.

Commercial Materials

Geoboards and rubber bands, unifix cubes, and magnetic numerals are welcome additions to any Math Center. If possible, these should be available to children at all times. A great selection of commercial games and manipulatives can also be found to supplement the handmade materials offered in this book. These can be obtained through catalogs, at conferences, and at teacher supply stores. Look for games that do not have a lot of pieces and can be played in five-to-ten minute intervals. Winning the game should not be emphasized. More appropriate expectations would include learning how to play the game, learning new math language (vocabulary), interacting appropriately with peers, and having fun.

Listening Activities

Listening is a very important skill, both academically and socially. In school, children need to learn to listen to the teacher and to take turns listening to each other. It takes practice and patience. The Listening Activities described in each unit offer children this needed practice.

Listening Activities are first presented to the whole group. Children must listen to the teacher, focusing on what he or she is saying and doing. Then, each child has a turn to answer a question or perform the demonstrated task. Initially, these questions/activities are simple ones—"What is your favorite color?" or "Choose your favorite color button and place it in the basket. Name the color you chose." The responses are not always limited to one correct answer, and the child should feel comfortable with his or her response. Gradually, math-related Listening Activities can be introduced. (*Example:* Count two paper clips and drop them into the container.) Model how the skill should be performed before asking the children to do the task.

Always model the appropriate way to answer a question. ("My favorite color is blue." "I made an AB pattern." "The fourth child is holding a soccer ball.") Encourage children to answer in complete sentences. Seasoned teachers may recognize the type of activities described above as "transition" activities. In a sense they are. They can be helpful when children are dismissed from one activity to another. However, the activities described in this book allow for much more than a transition. Children get to practice listening, taking turns, speaking in complete sentences, and of course, learning new math skills! Teachers benefit too. They get one-on-one opportunities to assess if a child is grasping the concepts that are being taught while at the same time building up the child's confidence. In addition, by focusing on one child at a time, the remaining children are given the opportunity to review the skill many times.

Children in any class will demonstrate a range of abilities. It is imperative to have a basic grasp of each child's abilities and tailor questions accordingly. Choose a child who has a good understanding of the skill to begin and let the less confident children watch and learn as the activity proceeds. Occasionally, a child will hesitate to answer. This is a good time to say something like, "I like the way Jenny is thinking about her answer." This takes some of the pressure off the child and rewards him or her for not rushing and guessing. Then, if necessary, the child can be guided to a successful outcome.

It is important for the teacher to respond to each child's response or skill with varied words of praise. For instance, when asked to complete a pattern the teacher might say, "I noticed that you were repeating the pattern to yourself before you added another piece. I like the way you were thinking about your answer. Good thinking today!" If the child's response is incorrect, assist the child in finding the correct answer. Follow this with another positive comment to build up the child's confidence.

Classroom manipulatives, children's literature, teacher-made materials, dry-erase boards, and the children themselves will all be used to carry out Listening Activities. It may be helpful to have a small tripod or easel available so that any pictures used can be displayed to make it easier for the children to see. This tripod can also be used to hold a flannelboard or dry-erase board.

This book contains many samples of good Listening Activities to get you started. Start an "ideas notebook" of your own. List ideas and supplies for quick and easy access. This notebook can be organized by math skills that are being taught. Develop some games of your own. If you find the children do not understand the concept, adjust accordingly so the children will succeed and have fun.

How to Use the Literature Selections

Be creative with the books you use with your children. Preview the selected books that have been listed for each unit and decide what will work best for your particular class. The books listed at the end of each unit should be available at a local bookstore or library. Additional suggestions have been added at the back of the book in the "Oldies but Goodies" section. This page lists books that may no longer be in print but can provide great additions to the math units. The local library, used bookstores, and websites are excellent resources that should not be overlooked. At the library, use the specific math goal described in this book as a "keyword" on a computer search. For instance, type in *pairs* and the preschool age group, and a list of books will appear that can be reviewed to see if there are appropriate titles for a particular class. Continue to add new books as they are published to build your Math Library.

Good literature can be incorporated in the classroom in a variety of ways:

- Children enjoy having books read to them during circle time or in small groups during free choice time. The books read by the teacher should always be made available for the children to look at ("read") again and again. Make sure a Reading Center is included in the classroom.

- Some of the books selected each month can be used effectively for Listening Activities. Once the book has been shared at group time, the teacher can use each page to ask the children different questions. Preview each book before using it to make sure there is enough information to prompt questions for each child. The book *How Many Snails?* by Paul Giganti (William Morrow & Company, Inc., 1994) is a good example of this type of book.

- Books can also be used to introduce a topic for discussion. The book *Just Enough Carrots* by Stuart Murphy (HarperTrophy, 1997) is a great book to use when introducing the concept of *equal*, *more*, and *less*. *Get Well, Good Knight* by Shelley Moore Thomas (Puffin Books, 2004) is a good book to introduce ordinal numbers.

- Books can be used for group guessing games. The book *How Many, How Many, How Many* by Rick Walton (Candlewick Press, 1996) is a series of riddles with number clues. Children will have fun guessing the answers in this book, which get progressively more difficult.

- Some books lend themselves to be looked at by one or two children at a time as an activity book. The book *Pigs from 1 to 10* by Arthur Geisert (Houghton Mifflin, 2002) is a seek-and-find book. The numerals 0–9 as well as 10 pigs are concealed on each page, accompanied by a catchy story.

- Other books can be added to the Writing Center to encourage children to copy numerals or words into their own work. The book *Let's Count* by Tana Hoban (Greenwillow Books, 1999) has a large numeral, the printed word, and a picture to illustrate each number.

Suggested Math Curriculum

Same and Different Sorting/Classifying/Sets Counting/One-to-One Correspondence The Calendar Left to Right 	Predicting and Estimating Patterns Graphing  Parts of a Whole/ What's Missing?	Ordinal Numbers Sequencing Seriation Time
Flat Shapes Solid Shapes 	Cardinal Numerals 0 – 10 Pairs Counting Backward 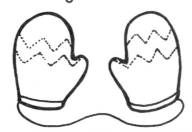	Measuring Equal, More, Less Money 
Addition Telephone Numbers	Subtraction Odd and Even 	Halves 

Same and Different

Educational Objectives: The child will be able to match items that are alike. The child will be able to verbalize what is the same or different between two or more items.

Vocabulary: *Same*—two items that are alike

Different—two items that are not alike

Classroom Activities

1. At circle time explain that you are going to talk about the words *same* and *different*. Review the definitions of these words and give examples. Call two children up to the front of the class. Have each child stand in a hula hoop (optional). Ask the children how they are the same/different. (*Examples:* both are girls; have shorts on; have brown hair; one has sandals on, the other has sneakers on; one is wearing glasses, the other is not; etc.) Continue until everyone has had a turn.

2. Call one child up to the front of the class. Explain that you are going to change something about the child to make him or her look different. Ask the children to look carefully at their classmate. Have them close their eyes. Change one thing about the child. (*Examples:* roll up a sleeve, remove glasses, remove a shoe, etc.) Have the children uncover their eyes and tell what is different.

3. Play a concentration game. Use a store-bought game or create one using the Abstract Shape Cards (pages 13–16). Make two sets of the cards and laminate them for durability. Cut the cards apart. Start with five pairs of cards and build up as skills improve. To play Concentration, turn the cards upside down and arrange them in rows. The first player turns over two cards so everyone can see them. If the cards match, then he or she keeps the pair and takes another turn. If they do not match, the cards are turned back over and the next player takes a turn. Players try to remember the location of specific cards. As more cards are revealed, it is easier for the players to find matching cards. Play continues until all the cards have been paired.

4. Use a collection of everyday items to encourage the children to do matching activities. Some examples could be socks, pairs of holiday napkins or plates and cups (Easter, Hanukkah, New Year's, Valentine's Day, etc.), earrings, or shoes.

Same and Different *(cont.)*

Listening Activities

1. Use the Same and Different Game Cards (pages 17–26). Copy the cards and glue them onto slightly larger construction paper to create color borders around the cards. Laminate them for durability. Explain to the children that each child will have a turn to look at a different card. He or she studies the card and discusses which picture on the card is different and why. Encourage the children to use complete sentences when answering.

2. Use the matching game Ready for School (pages 27–30) as a flannelboard activity. Copy the pages. Color the patterns on each child's clothes to match a corresponding book bag. Glue the cards onto construction paper and laminate them. Attach flannel to the back of each card. The children can use a flannelboard to match the clothes on the children with the matching book bags. This can be done independently, in small groups, or as a Listening Activity. Ask each child to find a book bag to match a shirt and explain to the group why it is a match. (*Example:* "The boy's shirt and the book bag both have blue stripes.")

3. Use the Boy and Girl Patterns (pages 31–32) to make a poster titled, "Same and Different." Color the pictures, making ten things the same and ten things different. Add the sentence, "What is the same and what is different about these children?" Show the poster to the class and explain that each child will have a turn to find something the same or different about the children. Encourage the children to use the words *same* or *different* as they describe their observations to the class. After this has been introduced as a Listening Activity, place the poster in the Math Center so the children can continue to find similarities and differences during free choice time.

Selected Literature

Blue's Clues Sort and Match (Learning Horizons, 2004)

Find and Fit Series by David Cook (Silver Dolphin, 2000)

A Pair of Socks by Stuart Murphy (HarperCollins Children's Books, 1996)

Shopping for School

2–4 Players

Developing Skill: Players will count the dots on the die to match the dots on the shopping list.

Materials

- 4 small baskets
- 4 scissors, 4 glue sticks, 4 pencils, 4 crayons, and 4 paintbrushes
- Shopping List on page 33 (copy and laminate for durability)
- 1 standard die (dots 1–6)
- large tray or basket

Playing the Game

1. Place all of the school tools in the large tray or basket. Place the shopping list in the center of the table for the players to share.
2. Each player takes a small basket.
3. Players take turns rolling the die and counting the number of dots showing.
4. Players match the number of dots on the die with the same number of dots on the shopping list. The player then collects the corresponding item from the tray or basket and puts it in his or her basket. Rolling a six (six dots) earns a free pick, and the player can select any item from the tray or basket that he or she may need.
5. Play continues until all players have obtained the five items on the shopping list.

Good Night Bear

2–4 Players

Developing Skill: Players will count the dots on the die to match the dots on the bear's pajamas and blanket.

Materials

- Bear Pattern (page 34)
- 4 sheets of brown construction paper (can be different shades)
- 4 different sheets of brightly colored construction paper
- 1 standard die (dots 1–6)

Preparation: Copy the bear pattern onto all eight sheets of construction paper and laminate. Cut the four brightly colored bear pages into six sections using the solid black lines as guides— pants, right-side top, left-side top, right slipper, left slipper, and blanket. Discard the leftover pieces.

Playing the Game

1. Each player chooses a bear to get ready for bed and a color of pajamas. He or she can collect the six colored parts needed and put them next to his or her bear.
2. Players take turns rolling the die and counting the number of dots showing.
3. Players match the number of dots on the die with the number of dots on the colored pajamas and blanket.
4. Play continues until all the bears are dressed and ready for bed.

Use with Classroom Activity #3 on page 10.

Abstract Shape Cards

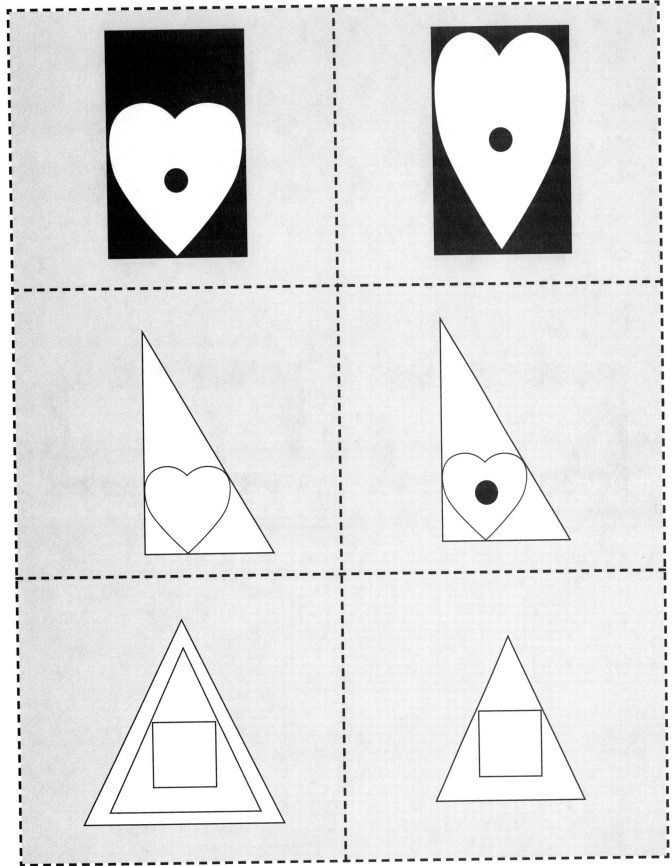

Abstract Shape Cards *(cont.)*

14

Use with Classroom Activity #3 on page 10.

Abstract Shape Cards *(cont.)*

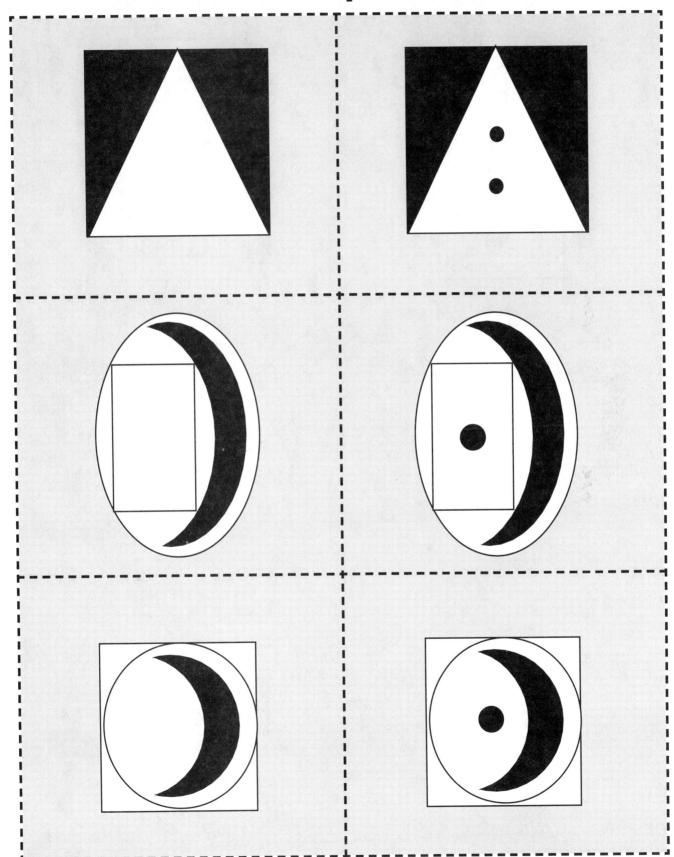

Abstract Shape Cards *(cont.)*

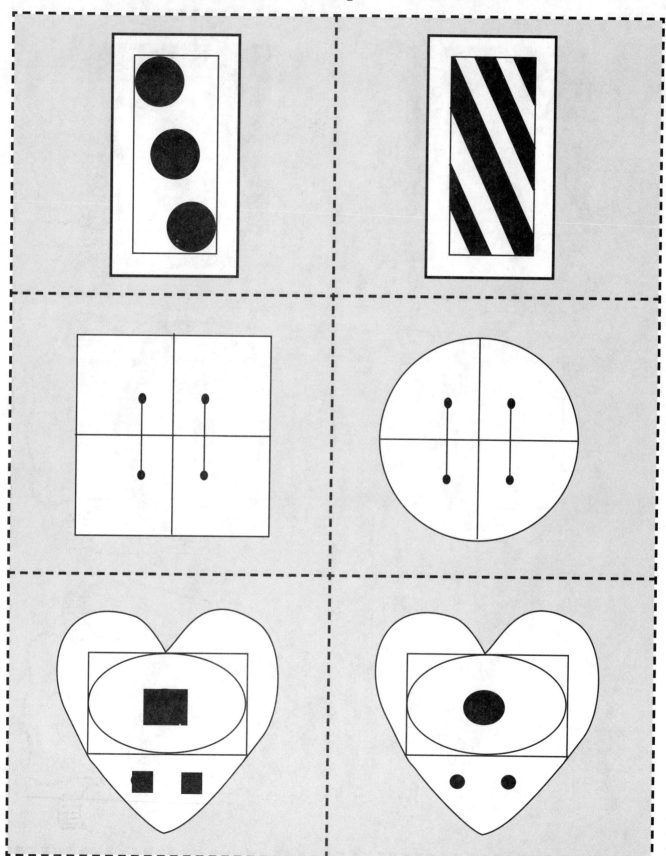

Use with Listening Activity #1 on page 11.

Same and Different Game Cards

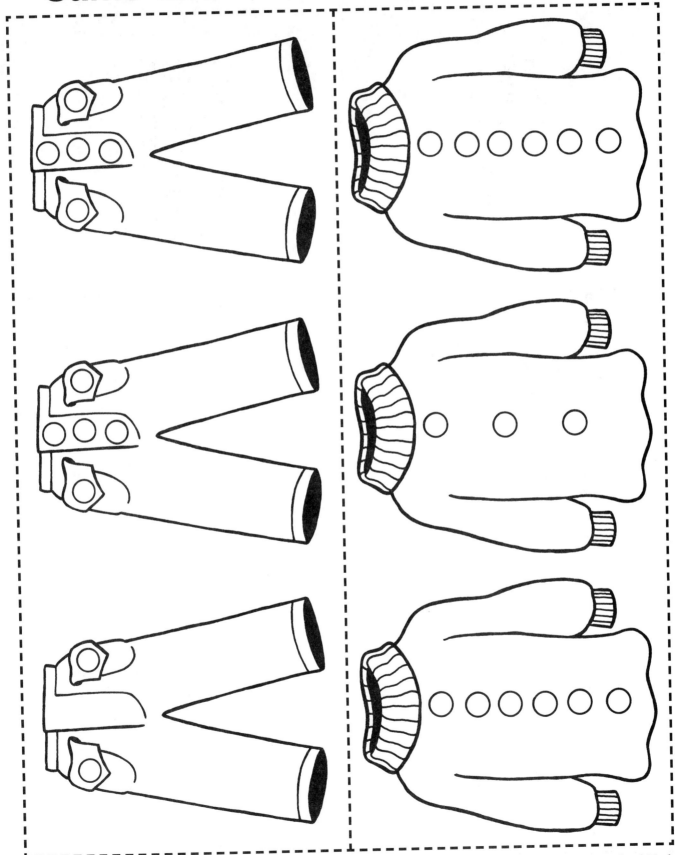

Same and Different Game Cards *(cont.)*

Use with Listening Activity #1 on page 11.

Same and Different Game Cards *(cont.)*

Same and Different Game Cards *(cont.)*

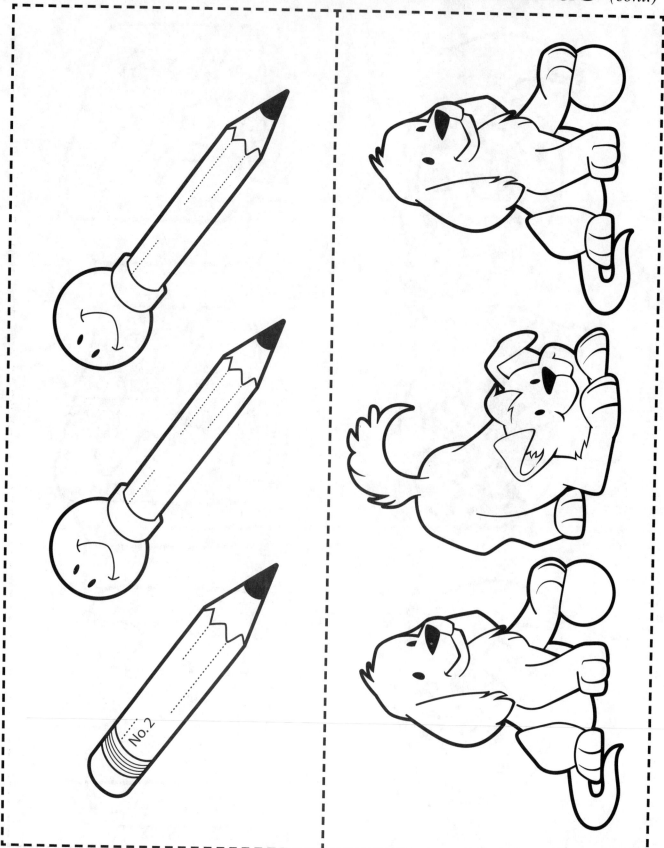

Use with Listening Activity #1 on page 11.

Same and Different Game Cards *(cont.)*

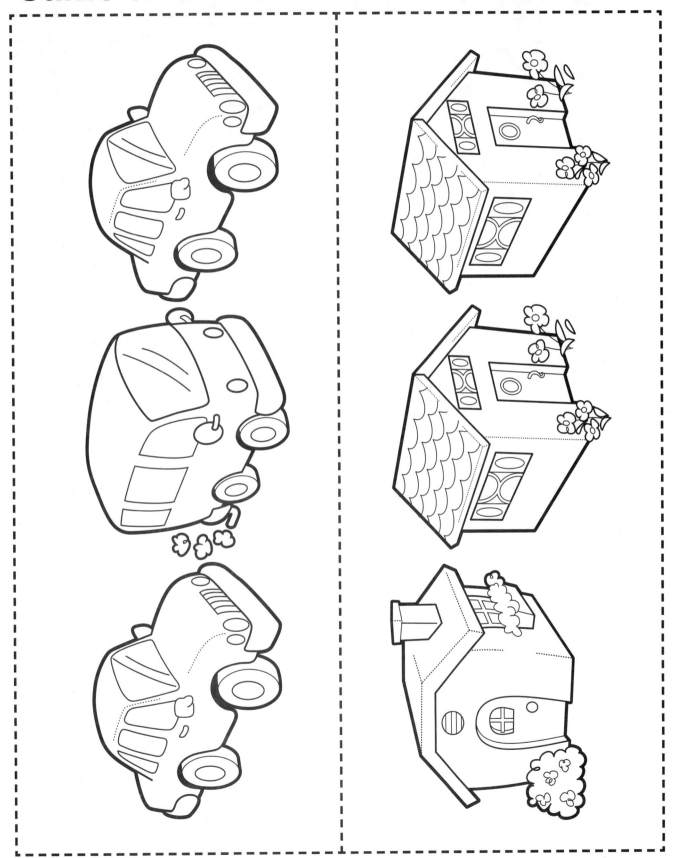

Same and Different Game Cards *(cont.)*

22

Use with Listening Activity #1 on page 11.

Same and Different Game Cards *(cont.)*

Same and Different Game Cards *(cont.)*

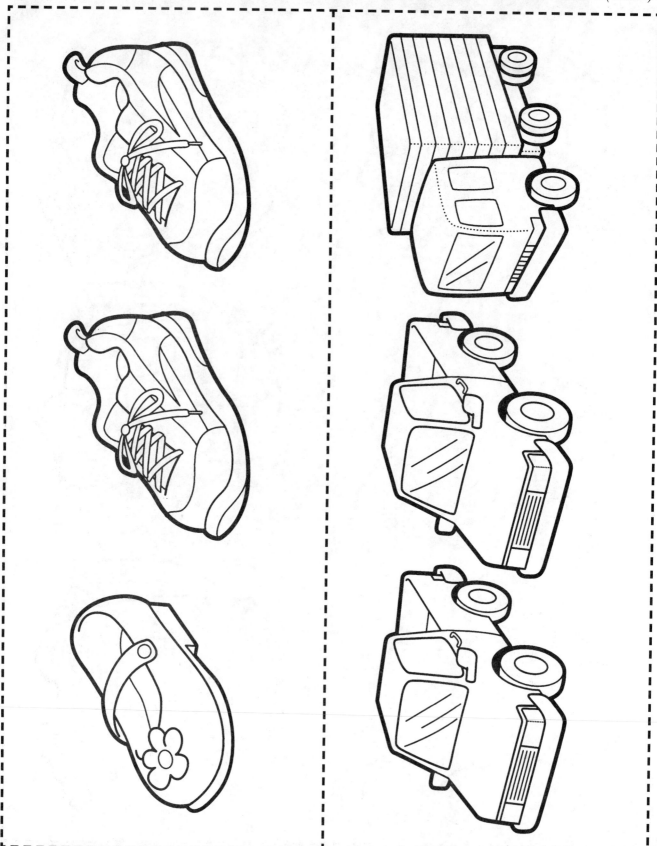

24

Use with Listening Activity #1 on page 11.

Same and Different Game Cards *(cont.)*

Same and Different Game Cards *(cont.)*

26

Use with Listening Activity #2 on page 11.

Ready for School

Ready for School *(cont.)*

Use with Listening Activity #2 on page 11.

Ready for School *(cont.)*

Ready for School (cont.)

30

Use with Listening Activity #3 on page 11.

Boy Pattern

Girl Pattern

Use with the Shopping for School game on page 12.

Shopping for School

scissors

glue stick

pencil

crayon

paintbrush

free pick

Bear Pattern

Sorting, Classifying, and Sets

Educational Objectives: The child will be able to group objects by common observable attributes (size, shape, color, pattern, position, use, etc.). He or she will be able to verbalize how the items are alike. (Classified by color, use, likes, etc.)

Vocabulary: *Sorting*—to arrange items according to a group

Classifying—to name a group according to a common bond

Set—a group of items that satisfies a given condition

Classroom Activities

1. At circle time explain that you are going to talk about sorting. Review the definitions. Work together to sort the types of shoes worn by each class member. Have each child take off one shoe and one at a time place it in front of the teacher. Point out something specific about each shoe as it is placed on the floor. Introduce sorting by demonstrating that there are a number of ways the shoes can be sorted and classified—by color, by closure (Velcro®, laces, slip on), by size, or by type (sneaker, sandal, hard sole). Suggest that the shoes need to be sorted into two groups (sets) and ask if the children have any suggestions as to how they should go about this task. After a decision has been made, bring out two hula hoops to use as visual tools to help the children understand sorting. Identify what kind of shoe should go in each hoop. One at a time, let each child retrieve his or her shoe and put it in the proper hoop. Analyze the results with the children. Use the word *set* when describing the two groups of shoes.

2. Explain to the children that they will be able to sort their snack as they eat it. Give each child a copy of Sorting Snack (page 38). Provide a snack of cereal, dried fruit, candy, and crackers. Place a handful of the snack on a napkin and let the children sort their snack as they enjoy it. Encourage discussion and ask questions as they sort.

3. Add a sorting poster to the Math Center. This can be purchased at a store or can be handmade. *Example:* A poster can show the different ways to sort foods (fruits, vegetables, milk products, meats, etc.) or types of animals (farm, zoo, pets, etc.).

4. Add commercial sorting toys to the Math Center such as vehicles, bugs, dinosaurs, etc.

5. Add sorting jars to the Math Center. Use plastic jars with different collections of items to sort. Provide a larger bin to hold two to three of these jars at a time. Demonstrate to the children how they can use mini loaf pans or muffin tins to sort the items. Model how to clean up and then sort another jar. *Examples:*

   Nature Jar: walnuts, buckeyes, pinecones, acorns
   Shop Jar: screws, nuts, bolts, wing nuts
   Ring Jar: assorted children's plastic rings
   Shell Jar: assorted small shells
   Eraser Jar: assortment of juvenile erasers and pencil-end erasers
   Golf Tee Jar: different colors, sizes, and materials
   Bottle Cap Jar: assorted sizes, types, and colors
   Button Jar: assorted sizes, colors, and number of holes
   Paper Clip Jar: different colors, sizes, and types

6. Send home a letter (page 37) asking your children to bring in a few sorting items to contribute to the Math Center. Children tend to participate more in the Math Center if they have ownership and have contributed to it. Change sorting items on a daily or weekly basis.

Sorting, Classifying, and Sets *(cont.)*

Listening Activities

1. Explain that each child will get a small selection of items and that he or she is to sort them into two sets. Hand out four to six items to each child. (Use the sorting items collected by the children and give a different selection to each child.) After he or she has completed the task, ask the child to explain to the class how he or she sorted the items.

2. Sort several items into groups in front of the class. Then, let each child reach into a bag or box for a new item. Ask him or her to put it in the correct group. Ask the child why it was put in a particular group. Encourage children to use complete sentences when answering.

3. Collect a large assortment of items that represents a particular season. Define three groups and use three hula hoops. (*Example:* Fall items could be classified as Halloween items, football items, and nature items.) Discuss the differences in the three groups. Some suggestions for items to collect would be a football helmet, jersey, padding; costumes, pumpkins, masks; and acorns, colored leaves, gourds. Explain to the children that they will each select one item and place it in the proper hoop according to how they decide to classify it. Each child should explain to the class his or her actions and why. This activity can lend itself to a discussion about how some items can be sorted into two different sets.

4. Use the What Doesn't Belong? Cards (pages 39–48). Explain to the children that each card illustrates a "set" of items and one thing does not belong in that set. Show one card to each student and have him or her tell the class which item does not belong in the set and why. Encourage children to use complete sentences when answering. Model how to do this.

5. Introduce the auditory version of the What Doesn't Belong? game. Explain to the children that this is an activity where they have to listen to what word does not belong in a set.

What Doesn't Belong? (Auditory Version)

Directions: Read one grouping at a time and have a child tell you which item does not belong in the set and why. This is a verbal game that helps develop auditory skills.

1. ball, cat, bat
2. salt, tiger, pepper
3. brush, comb, hammer
4. table, coat, chair
5. mud, pencil, paper
6. cat, pizza, juice
7. necklace, bracelet, cup
8. pencil, coat, hat
9. whale, bread, butter
10. car, train, milk
11. first, tomato, third
12. Monday, elephant, Sunday
13. strawberry, boat, peach
14. happy, angry, cupcake
15. triangle, square, apple
16. blue, horse, orange
17. shirt, snake, pants
18. monkey, shoe, sock
19. mountain, ocean, shirt
20. three, mushroom, eight

Selected Literature

The Button Box by Margarette Reid (Puffin Books, 1995)

Dave's Down-to-Earth Rock Shop by Stuart Murphy (HarperTrophy, 2000)

A First Book About Mixing and Matching by Nicola Tuxworth (Gareth Stevens Publishing, 1999)

Gray Rabbit's Odd One Out by Alan Baker (Kingfisher, 1999)

Let's Sort by David Bauer (Yellow Umbrella Books, 2003)

Sets: Sorting into Groups by Michele Koomen (Bridgestone Books, 2001)

Sorting and Sets by Henry Pluckrose (Gareth Stevens Publishing, 2001)

Sorting Foods by Patricia Whitehouse (Heinemann Library, 2002)

Three Little Firefighters by Stuart Murphy (HarperCollins Children's Books, 2003)

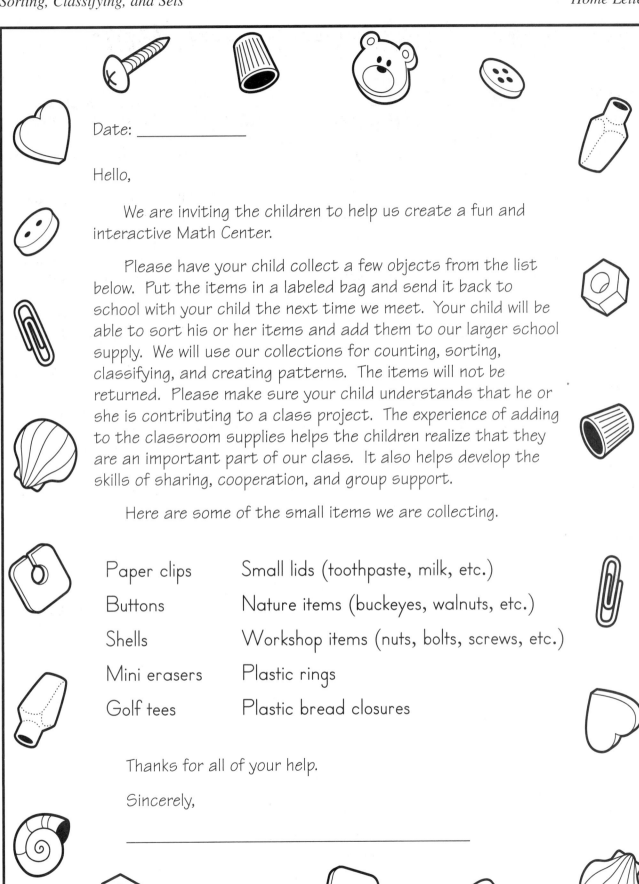

Date: _____

Hello,

   We are inviting the children to help us create a fun and interactive Math Center.

   Please have your child collect a few objects from the list below. Put the items in a labeled bag and send it back to school with your child the next time we meet. Your child will be able to sort his or her items and add them to our larger school supply. We will use our collections for counting, sorting, classifying, and creating patterns. The items will not be returned. Please make sure your child understands that he or she is contributing to a class project. The experience of adding to the classroom supplies helps the children realize that they are an important part of our class. It also helps develop the skills of sharing, cooperation, and group support.

   Here are some of the small items we are collecting.

Paper clips	Small lids (toothpaste, milk, etc.)
Buttons	Nature items (buckeyes, walnuts, etc.)
Shells	Workshop items (nuts, bolts, screws, etc.)
Mini erasers	Plastic rings
Golf tees	Plastic bread closures

   Thanks for all of your help.

   Sincerely,

   _____

Sorting Snack

I can sort my snack.

Fruit

Cereal

Candy

Crackers

What Doesn't Belong? Cards

Sorting, Classifying, and Sets

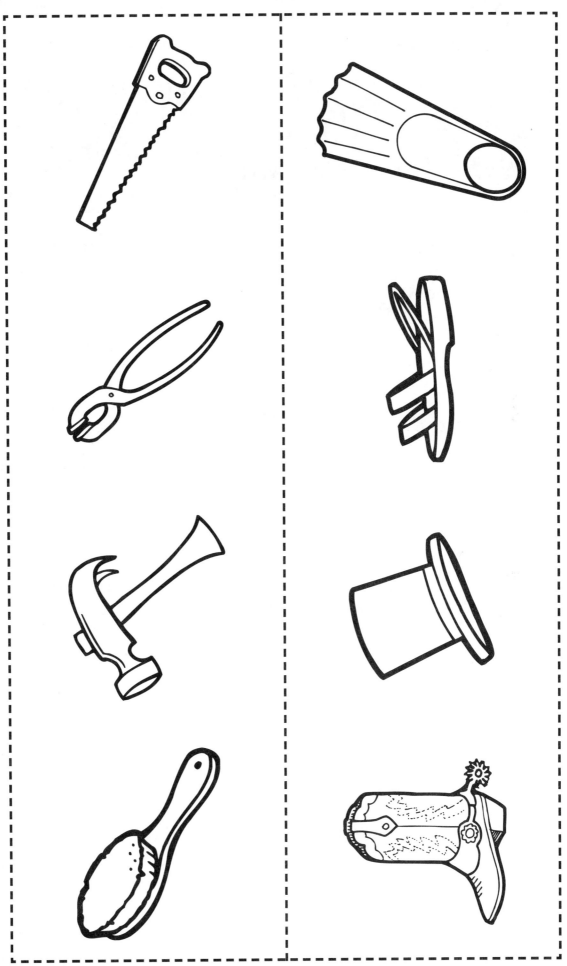

Sorting, Classifying, and Sets

What Doesn't Belong? Cards *(cont.)*

What Doesn't Belong? Cards *(cont.)*

Sorting, Classifying, and Sets

Sorting, Classifying, and Sets

What Doesn't Belong? Cards *(cont.)*

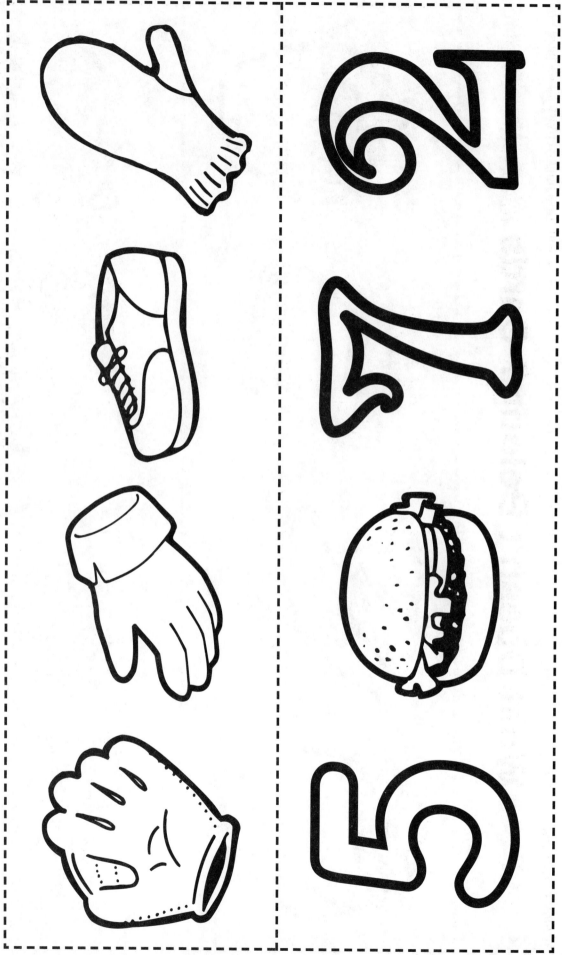

What Doesn't Belong? Cards *(cont.)*

Sorting, Classifying, and Sets

43

What Doesn't Belong? Cards *(cont.)*

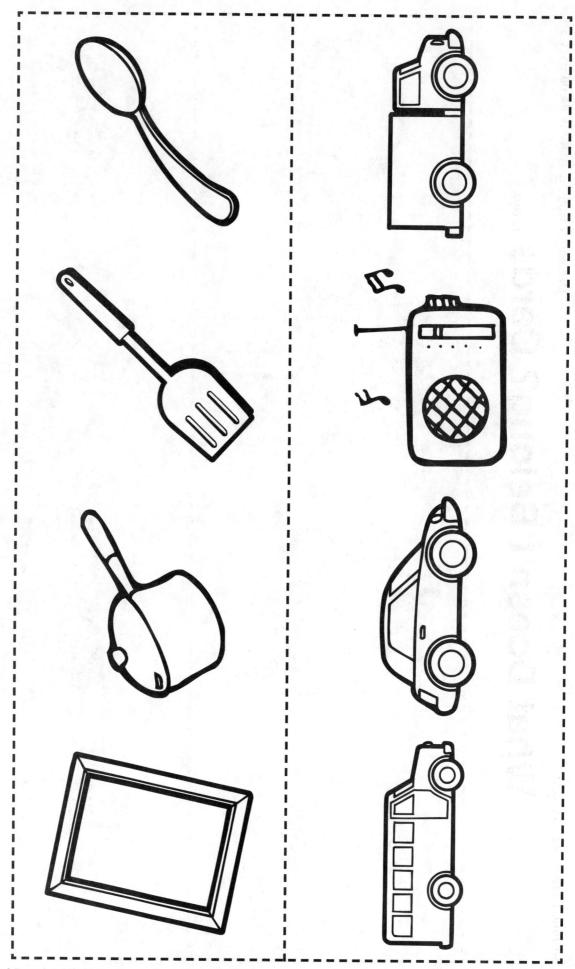

What Doesn't Belong? Cards *(cont.)*

Sorting, Classifying, and Sets

What Doesn't Belong? Cards *(cont.)*

Sorting, Classifying, and Sets

46

What Doesn't Belong? Cards (cont.)

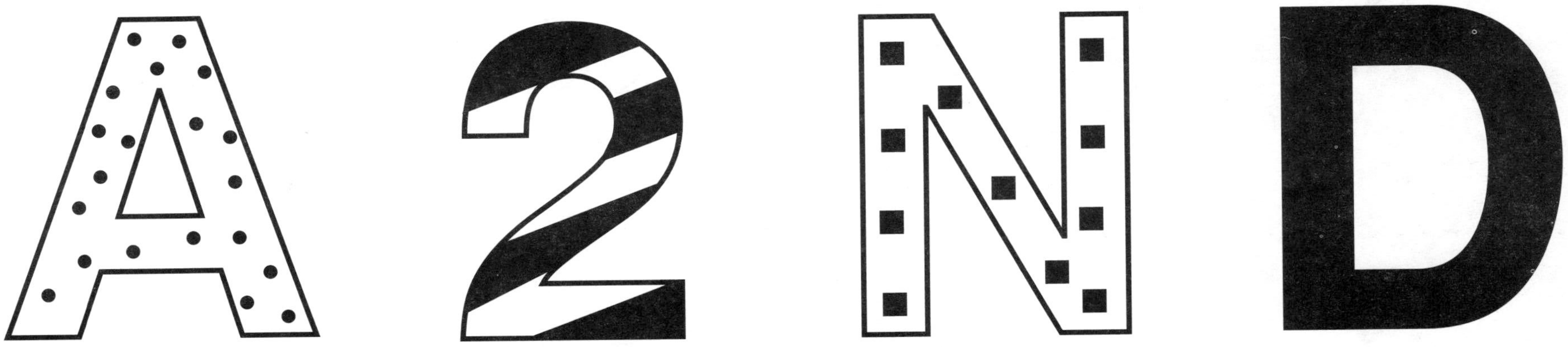

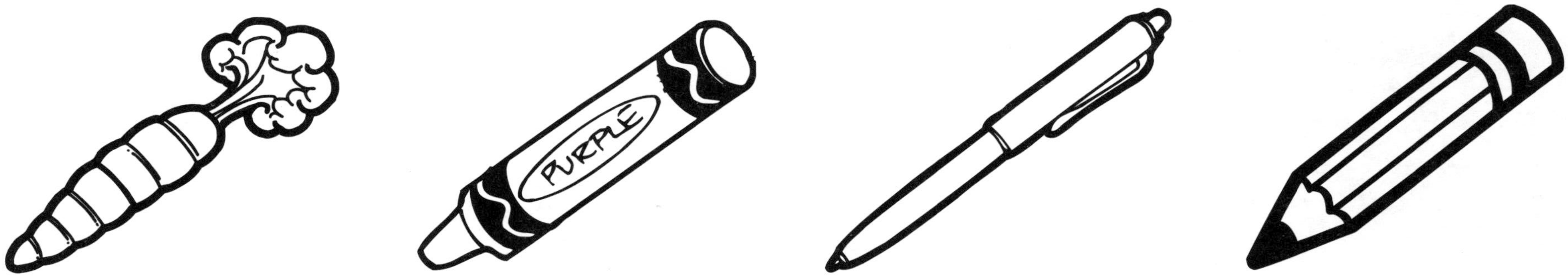

D N 2 A

PURPLE

What Doesn't Belong? Cards *(cont.)*

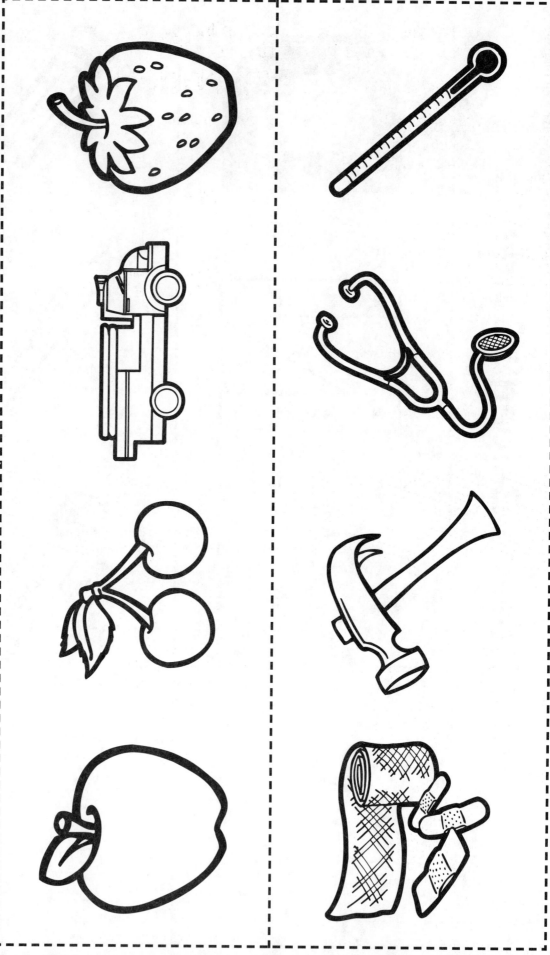

Counting, One-to-One Correspondence

Educational Objectives: The child will be able to accurately count 10 items by touching each item as it is counted.

Vocabulary: *Counting*—to name numbers in regular order

One-to-One Correspondence—accurately coordinating number names with objects being counted

Classroom Activities

1. Each day at calendar time, point to and count the numbers on the calendar. When the children have a good grasp of this skill, let a different child do this task each day. It is helpful to provide the child with a novelty pointer to make this job even more special and to assist him or her with the task. Using a pointer helps the child count using one-to-one correspondence and helps the other children see the number as they hear it.

2. Use different methods of counting the children when taking attendance to ascertain that they are all present. Pointing, tapping, and asking each child to sit down when he or she has been counted are all visual ways to count the class and model one-to-one correspondence.

3. Teach the children to count off to see how many children are at school. (Each child says one number until all the classmates have been counted.) Initially, the teacher will need to point to each child to prompt him or her to say a number. Eventually, the children should be able to do it independently after the teacher indicates who should start the counting.

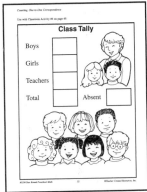

4. Fill in a Class Tally each day (page 52). After the number of children present has been established, ask the children how many classmates are absent. Next, find out how many teachers are present. This chart can be laminated and filled in with an erasable pen each day. (*Note:* You may wish to place your class picture over the generic one provided.)

5. Copy pages 237–240 and glue them together to make one long number line. Book binding tape or contact paper can be used to attach it to the table in the Math Center if the table is long enough. After this has been prepared, gather the class around the number line. This is a good time to use the items in the sorting jars that have been brought in by the children. Hold up a jar and ask the class to estimate how many items they think are in the jar. Demonstrate how to use the number line by putting one item on top of each numeral. Keep the number of items in the jars to a number that the children will be able to work with successfully. Individual number lines can also be made on long pieces of construction paper or adding machine tape which can be laminated. These can be used for group activities.

6. Count while glue is drying on a craft project. Use this activity throughout the year by expanding the number. Have students count to 5, 10, 20, etc. Younger students can count to five or ten several times.

7. Encourage the children to count objects in the school. (*Examples:* How many steps on a staircase? How many doors they pass to get to the classroom? How many bulletin boards are in the hall?)

8. At snack time, let children select one of each item offered for a snack. (*Example:* one cracker, one carrot, and one grape, or let a child pass out snacks so each child gets one of each item.)

Counting, One-to-One Correspondence *(cont.)*

Listening Activities

1. Explain that each child will be given a number. He or she is to count aloud to that number only and then stop. (*Variation:* Ask the child to count as he or she jumps the required number.)
2. Give a child a random number up to 10. Ask him or her to count that number of classmates.
3. Use any small item for a Keep Counting game. The teacher counts out the first couple of number items and then ask a child to finish counting them up to a certain number.
4. Use any fun collection of manipulatives and ask each child to count out a certain number.
5. The teacher chooses a number at random. The children then do a count off. When they get to the chosen number, that child is excused. Repeat until all children have been excused.

Selected Literature

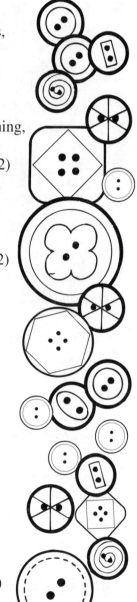

Anno's Counting Book by Mitsumasa Anno (HarpersTrophy Children's Books, 1986)

Blast-Off! A Space Counting Book by Norma Cole (Charlesbridge Publishing, 1994)

The Cheerios Counting Book by Barbara McGrath (Scholastic, Inc., 1998)

Chicka Chicka 123 by Bill Martin, Jr. (Simon and Schuster Children's Publishing, 2004)

Count and See by Tana Hoban (Simon & Schuster Children's Publishing, 1972)

Counting Animals by Cynthia Cappetta (Innovative Kids, 2001)

How Many Feet? How Many Tails? by Marilyn Burns (Scholastic, Inc, 1996)

How Many, How Many, How Many by Rick Walton (Candlewick Press, 1996)

How Many Snails? by Paul Giganti (HarperTrophy, 1994)

The Icky Bug Counting Book by Jerry Pallotta (Charlesbridge Publishing, 1992)

Let's Count by Tana Hoban (Greenwillow Books, 1999)

Math Fables by Greg Tang (Scholastic, Inc., 2004)

Math for All Seasons by Gregory Tang (Scholastic, Inc., 2002)

Monster Math by Anne Miranda (Voyager Books, 2002)

Mouse Count by Ellen Walsh (Harcourt, 1995)

My First Abacus Book by Nick Sharratt (Voyager Books, 1995)

The Napping House by Audrey Wood (Harcourt, 1984)

One Duck Stuck by Phyllis Root (Candlewick Press, 1998)

One Happy Classroom by Charnan Simon (Children's Press, 1997)

One, Two, Skip a Few by Roberta Arenson (Barefoot Books, 2000)

One Witch by Laura Leuck (Walker and Company, 2003)

Only One by Marc Harshman (Dutton Children's Books, 1993)

Spots: Counting Creatures from Sky to Sea by Laura Regan (Harcourt, 1999)

Teeth, Tails, and Tentacles by Christopher Wormell (Running Press Book Publishers, 2004)

Ten Black Dots by Donald Crews (William Morrow & Co., 1995)

12 Ways to Get to 11 by Eve Merriam (Center for Applied Research Ed., 1998)

Two Peas in a Pod by Annegert Fuchshuber (Lerner Publishing Group, 1998)

The Very Hungry Caterpillar by Eric Carle (Philomel, 1994)

Four in a Row   **2 players**

Developing Skill: Players will need to be able to count four markers in a row to win.

Materials

- Gameboard (page 53)
- 20–30 small novelty erasers (10–15 of two different kinds)

Preparation: Glue the gameboard onto construction paper and laminate it for durability.

Playing the Game

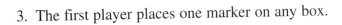

1. The object is to get four markers in a row.

2. Each player selects which set of markers (erasers) he or she will use.

3. The first player places one marker on any box.

4. The next player can place a marker anywhere on the grid, but he or she also wants to keep the other player from getting four markers in a row. Play continues until someone gets four in a row.

Classroom Count   **2 or more players**

Developing Skill: Players will need to be able to count various objects in the classroom.

Materials

- list of counting questions

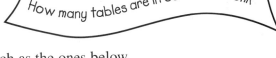

How many tables are in our classroom?

Playing the Game

Players take turns asking and answering questions such as the ones below.

- How many tables are in our classroom?
- How many clocks are in our classroom?

- How many windows are in our classroom?
- How many 2s are on the calendar?

- How many flags are in our classroom?
- How many doors are in our classroom?

- How many students are in our room today?
- How many teachers are in our room now?

- How many girls are here today?
- How many boys are here today?

Use with Classroom Activity #4 on page 49.

Class Tally

Boys

Girls

Teachers

Total

Absent

Use the gameboard with the Four in a Row game on page 51.

Four in a Row

Calendar

Educational Objectives: The child will recognize that the calendar is a tool used to measure time in terms of days, weeks, months, and years. He or she will be able to say the days of the week and be able to answer questions about yesterday, today, tomorrow, the day, month, and the year with the assistance of a calendar.

Vocabulary: *Calendar*—a tool used to measure time by days, weeks, months, and years

Yesterday—the day before today

Today—the present time or period

Tomorrow—the day after today

Classroom Activities

1. Review all the parts of the calendar, including the day of the week, number, month, and year each day at calendar time. Point out that each day ends with the letters *day*. Highlight these letters with a bright color. Explain that every day of the week has this same ending. If someone asks what day it is, the answer should have the word part *day* in it. This helps the children identify the days on the calendar.

2. Use a poem, chant, or song to help the children learn the days of the week and the months of the year. The cassette/CD "Dr. Jean and Friends" (Progressive Music, Tampa, Florida) has some easy and fun songs to help with these skills.

3. If the children only come to school some of the days of the week (e.g., Monday, Wednesday, Friday), ask them often what days they come to school to reinforce using the words in their vocabulary. This same teaching aid can be used if they have some kind of special activity only one day of the week.

4. Older children are ready to learn the concepts of yesterday, today, and tomorrow. Make a pocket chart using tagboard and attach it to the classroom calendar. (See diagram.) Glue the Days of the Week Cards (pages 56–57) onto colorful construction paper and laminate them. Each day change the cards accordingly and put them in the appropriate pockets on the chart. (Hide the leftover cards behind the three that are being used so they will be easily available to change on the next day). At calendar time, review what day it is and add the appropriate name card to the Today pocket. Then, demonstrate how to find out what day was yesterday and what day will be tomorrow by looking on the calendar. Review this concept each day to help the children get a better understanding of these terms. It is fun to let a special child fill in the pocket chart every day. Use the words *yesterday, today,* and *tomorrow* in everyday language to help model these words for the children.

Sunday	Monday	Tuesday
Yesterday	Today	Tomorrrow

Calendar *(cont.)*

Listening Activities

1. Recite the days of the week and stop at different places. Ask a child what day comes next. Have a calendar available as a visual tool to assist the children.

2. Ask each child a different question about the calendar. For example, ask a child to come up and point to and say a day of the week, the month, or the year on the calendar. Ask which day comes before or after another day or how many days are in a week. Encourage children to use complete sentences when answering.

Selected Literature

All Through the Week with Cat and Dog by Rozanne Williams (Sagebrush, 2002)

Big Week for Little Mouse by Eugenie Fernandes (Kids Can Press, 2004)

Clifford's Busy Week by Norman Bridwell (Scholastic, Inc., 2002)

Cookie's Week by Cindy Ward (Penguin Putnam, 1997)

Dear Daisy, Get Well Soon by Maggie Smith (Sagebrush, 2003)

Diary of a Wombat by Jackie French (Clarion Books, 2003)

Max and Ruby's Busy Week by Rosemary Wells (Grossett & Dunlap, 2002)

My Week by Josie Firmin (Candlewick Press, 2001)

One Lighthouse, One Moon by Anita Lobel (Greenwillow Books, 2000)

Pepper's Journal by Stuart Murphy (HarperTrophy, 2000)

Ruby the Copycat by Peggy Rathman (Scholastic, Inc, 1993)

Today Is Monday by Eric Carle (Philomel Books, 1983)

The Very Hungry Caterpillar by Eric Carle (Philomel Books, 1994)

A Week by Robin Nelson (Lerner Publications, 2001)

Days of the Week Cards

Sunday	Monday
Tuesday	Wednesday

Use with Classroom Activity #4 on page 54.

Days of the Week Cards *(cont.)*

# Thursday	# Friday
# Saturday	# Today
	# Tomorrow
	# Yesterday

Left to Right

Educational Objectives: The child will be exposed to the concept that the presentation of words, sentences, numerals, calendars, sequencing, seriation, etc., moves from left to right. (Although this concept is usually presented as a language arts goal, it applies to math and should be emphasized often.)

Vocabulary: *Left*—all or part of the left side

Right—all or part of the right side

Classroom Activities

1. Introduce the left-to-right concept to the children by starting to read a big book backward from the last page and from right to left. (It works best if you use a book with only one sentence on each page.) See if the children pick up on this confusion. If not, stop and declare that something is not RIGHT! Explain that words and numerals are read from left to right. Point out this concept within the classroom on the calendar, book titles, page numbers in a book, numbers on the number chart, etc. Then read the book the correct way and emphasize that it now makes sense. Review with the children that numerals make sense if they are read from left to right. This concept should be emphasized throughout the year so the children will be comfortable with this skill as they progress with their math, reading, and writing lessons.

2. Ask the children to work from left to right when presented with an activity that should be completed in this manner. This often will mean that at circle time (when the teacher is facing the children) the teacher will need to model these types of activities backward so the children are always seeing new information presented in the proper format, *left to right*. Remember to repeat this skill and the words *left to right* throughout the year when a new math concept is introduced that should be performed in this manner.

3. On the day the left-to-right concept is introduced, put an "L" stamp on each child's left hand to remind him or her of the left side of the body. If "L" stamps are unavailable, a sticker on the left hand will do.

4. Make up a silly poem, chant, or song to help the children remember to do their work from left to right and from top to bottom. Once the idea has been presented, remember to use it throughout the year to help children reinforce this important concept. (*Examples:* "Left side, left side. Now I know. That is where my pencil goes." or "Start on the left and move to the right. Now I know my work will be right.") Be creative and come up with a chant that works with your class.

5. Let children practice this concept by dipping small cars into shallow pans of paint and driving the cars from the left side of the paper to the right. It helps to place the word *left* on the left side of the paper and *right* on the right side.

6. Use the Left-to-Right Practice sheets (pages 60–61) to encourage the children to move a crayon or marker from left to right. Make several copies of the sheets, glue them onto colorful paper, and laminate them. Let the students use dry-erase markers or oil pastel type crayons that erase easily with a tissue. (Make sure the children position the paper correctly in front of them and that they do not rotate the papers 90 degrees to complete some of the strokes. This will ensure they have a true experience of going from left to right.)

Left to Right *(cont.)*

Listening Activities: (**Note:** All listening games that are used throughout the year should be completed using the left-to-right format if it is appropriate for the specific task.)

1. Use the Left-to-Right Cards (pages 62–71). Explain to the children that they will look at pictures of three animals. On each card two of the animals are traveling from the left to the right, but one is not going the correct way. Ask each child to tell you which animal is not moving from the left to right.

2. On a large piece of paper write the word *left* on the left side and *right* on the right side. Explain to the children that they are going to work together to make a colorful left-to-right poster. Let each child come up and choose a marker and draw a line from the left side of the paper to the right. The lines can be wiggly, zigzag, curvy, etc., but they need to go from left to right. Lines can go on top of other lines as well. Ask each child to tell the class what color he or she chose and what kind of line he or she will add to the poster. Display this unusual artwork for the children to enjoy.

3. Call on one child at a time to come up and arrange a specific number of items in order from left to right. (*Example:* "Can you line up three bottle caps on this line going from left to right?") For this activity, it is fun to use some of the items the children have brought in to sort. A bright piece of paper with a bold line drawn on it helps the children place the items from left to right. Have the child repeat the Left-to-Right Chant that has been taught, or say "left to right" as the items are being put on the line.

4. Ask each child a question about items that are read from left to right. (This list will need to be specific to the classroom that is being used.) For example: "When counting the numbers on the calendar, where do I start?" or "When I read the year on the calendar, what number do I look at first?" or "When I read the page numbers in a book, where do I start?" or "When I write my numbers in order, where do I start?" All the answers should be *left* so the children get the idea to always start on the left. As each child responds with the word *left*, it is helpful for the teacher to demonstrate the skill to reinforce this concept.

Selected Literature

Bear's Left and Right by Keith Faulkner (Dutton Children's Books, 2002)

A Fly Went By by Mike McClintock (Random House Books for Young Readers, 1986)

Which Way, Ben Bunny? by Mavis Smith (Scholastic, Inc., 1996)

Left-to-Right Practice

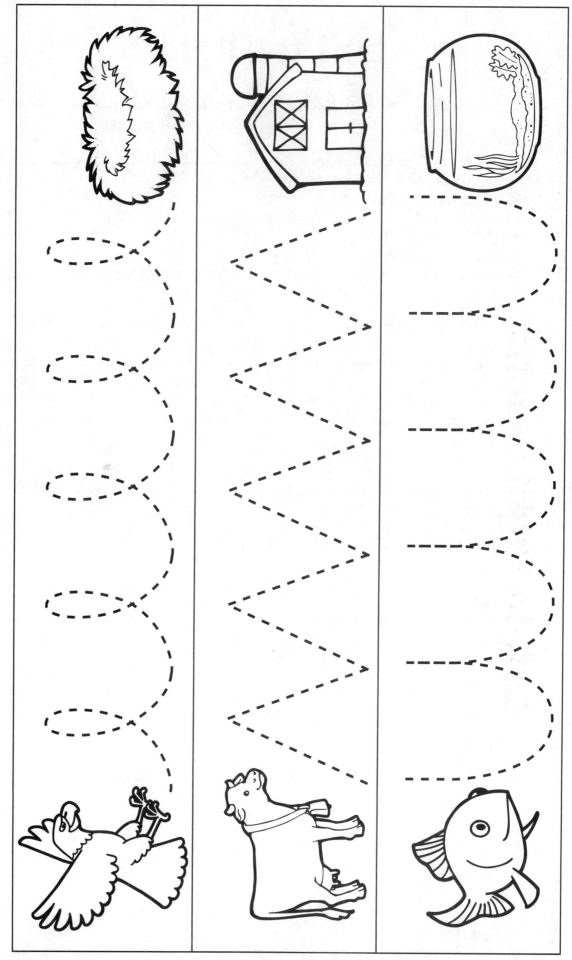

Left-to-Right Practice (cont.)

Left to Right

Left-to-Right Cards

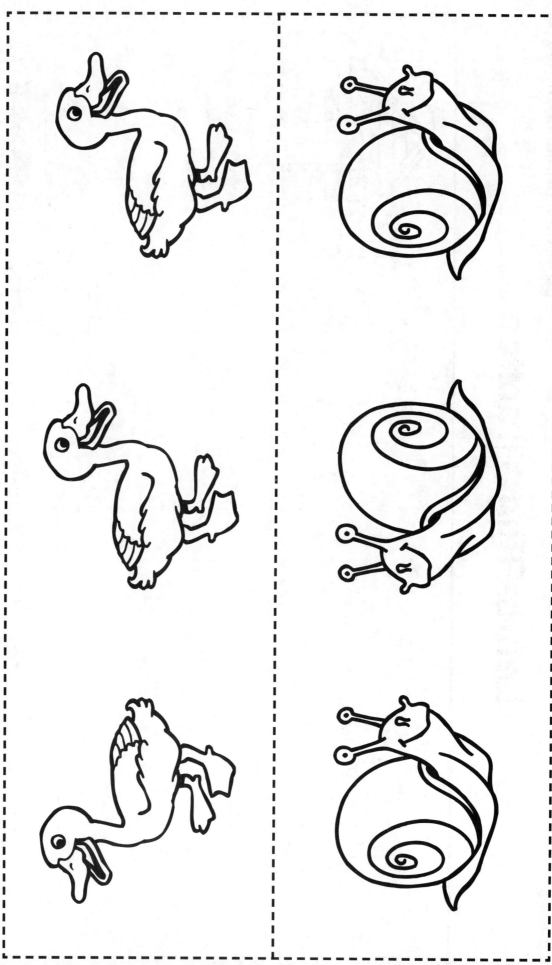

Left-to-Right Cards *(cont.)*

Left to Right

Left-to-Right Cards *(cont.)*

Left to Right

Left-to-Right Cards *(cont.)*

Left to Right

Left-to-Right Cards *(cont.)*

Left to Right

Left-to-Right Cards *(cont.)*

Left-to-Right Cards *(cont.)*

Left to Right

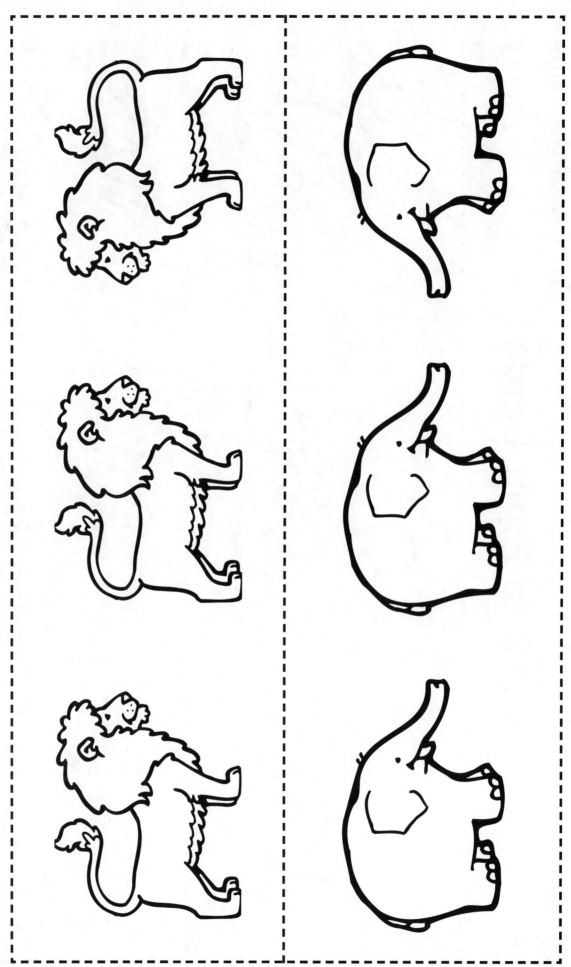

68

Left-to-Right Cards *(cont.)*

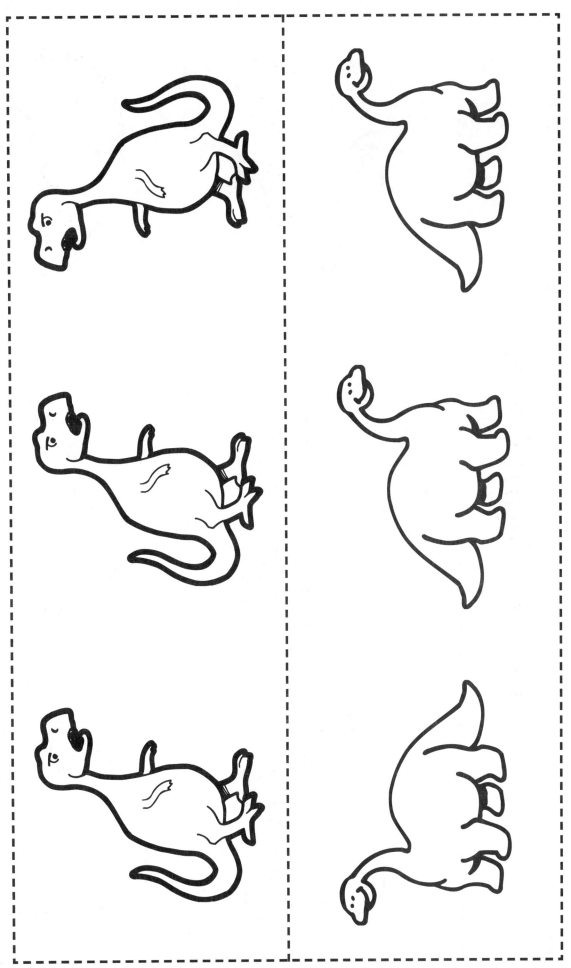

Left to Right

Left-to-Right Cards *(cont.)*

Left to Right

Use with Listening Activity #1 on page 59.

Left-to-Right Cards *(cont.)*

Left to Right

Predicting and Estimating

Educational Objectives: The child will be able to look at a presented stimulus, think about the possibilities, and then verbalize in advance what he or she thinks will happen in a given situation.

Vocabulary: *Predict*—to say in advance what one believes will happen

                 Estimate—to calculate how much of an unknown item there is, approximately but carefully (size, value, cost, etc.)

Classroom Activities

1. Introduce predicting by asking the children to predict the outcome of an action taken by the teacher. *Example:* Pour a container of cube blocks into a container a little smaller than the original one. Explain that this skill involves three steps.

        First—look at the situation.

        Second—think about the possibilities.

        Third—verbalize what will happen in a given situation.

   Invite discussion with the children to see what they predict will happen. Use the We Predict chart (page 74) to record the children's thoughts. Explain that the only way to find out what will happen is to do the action. Record what actually did happen. Display the chart for parents to enjoy.

2. Introduce estimating by explaining that this skill is similar to predicting, but it often involves numbers. Provide an interesting item. (*Examples:* a basket of gourds, a bowl of leaves, or a bag of plastic Halloween rings.) Ask the children to estimate how many items are in the container. Once again the children must consider three steps. First—look at the item. Second— think about the possibilities. Third—calculate an answer. (The teacher can talk about an unrealistic  number to make a point and help the children learn about this skill.) Use the Let's Estimate chart (page 75) to record each child's guess. After every child has given his or her estimate, point out that the only way to find the actual number is to count the items. Have the class count aloud with the teacher to find the final answer. Record the actual number on the chart. Display the chart for parents to see.

3. At snack time, encourage your students to estimate how many pieces of snack are in a snack bag. Remind them to count each item to find out the actual number.

4. Return to predicting and estimating activities several times throughout the year. It only takes a short amount of time, and it is a good way to introduce something new. Letting the children predict what they think will happen is a good way to start many science experiments. Keep extra copies of the estimating and predicting activities for those spontaneous teaching moments in the classroom.

Predicting and Estimating *(cont.)*

Classroom Activities *(cont.)*

5. At circle time, introduce a prediction activity. Provide assorted sizes of Valentine candy boxes to be used with play dough. Make a batch of brown chocolate play dough. (It is fun to actually put cocoa powder in the play dough if you are sure the children will not eat the dough.) Show the children how to roll balls to simulate chocolate candy pieces. Ask them to predict how many chocolate candies they think will fit in one of the boxes. Complete the task and ask, "How many pieces actually fit in each box?" Explain that the children can use this information to predict how many candies they can get in the other boxes. Did they get more or less than their predictions? Ask the children to share their results with you. (It is interesting to see how some of the children will adjust the size of their candy balls so that their prediction will come out correct. This is a good problem-solving skill.)

6. When reading a good book with a surprise ending, stop and ask the students what they predict will happen. It may be fun to write down their predictions and talk about them before finishing the book. The book *Caps for Sale* by Esphyr Slobodkina is a good book to use for this activity. Ask the children what they think the peddler should do to get his caps back.

Listening Activities

1. Follow up on the estimating activity introduced at circle time by showing the children a different plastic see-through bag with an unidentified number of items in it. Let each child estimate how many total items are represented. Record each child's guess on the Let's Estimate sheet. After everyone has recorded his or her answer, count the items in the bag to find out the actual amount. Display the results where parents and children can review them.

2. Gather an assortment of any small sorting items or toys in a large bowl. Introduce a small scoop, such as the kind found in laundry detergents, or a ½ or smaller cup measure. Explain that each child will have one turn to scoop out some of the items. Ask each child to predict if he or she will be able to scoop out a specific number of items in one try. Use the prediction chart and record each child's prediction before he or she takes a turn. Let each child try this activity and record the results. Display the results for parents to enjoy.

Selected Literature

Caps for Sale by Esphyr Slobodkina (HarperTrophy, 1987)

Estimating: How Many Gollywomples? by John Burstein
(Weekly Reader Early Learning Library, 2003)

Geoffrey Groundhog Predicts the Weather by Bruce Koscielniak (Houghton Mifflin, 1998)

The Little Old Lady Who Was Not Afraid of Anything by Linda Williams (HarperTrophy, 1998)

Use with Classroom Activity #1 on page 72.

We Predict

The action: _____.

What we think will happen: _____.

Child's name	Prediction

What actually happened? _____

_____.

Use with Classroom Activity #2 on page 72.

Let's Estimate

How many _____**?**

Child's name	Estimate

Actual number:

Patterns

Educational Objectives: The child will be able to identify a pattern, label a pattern using the ABC terminology (e.g., ABAB, ABCABC, ABACABAC), and complete a pattern that has been started by the teacher. The child will also be able to independently create a simple pattern.

Vocabulary: *Pattern*—to make, do, shape, or plan in imitation of a model
Identify—to recognize as being the thing described
Label—a descriptive word or phrase applied to a group or object
Complete—to end, finish, or conclude

Classroom Activities

1. Introduce the concept of patterns with a simple AB pattern using two small toys of different colors. Explain that the pattern can be identified using descriptive words by saying the colors red, green, red, green, or by using ABC language and saying A, B, A, B. Start a pattern and ask what will come next. Ask children to identify the pattern using ABC language. Tell the children that during the year they will be learning several new patterns and using the ABC language that identifies the different patterns. Introduce new patterns when you are sure the children have a good understanding of the initial one. Use the ABC language often. You will know you have had success when a child spontaneously proclaims, "I made an ABC pattern!"

2. Help the children get a visual picture of patterns. For example, have the children arrange themselves to create patterns such as boy/girl; standing/sitting; hands up/hands down; etc. Start with AB patterns and progress to more difficult ones.

3. Use the classroom calendar numbers to visually display a different color pattern each month. In September, begin with an AB pattern. In October, use ABC; November, AAB; December, AABB; etc. (Buy or make at least three different sets of colored numbers.)

4. Patterns can also be auditory. Introduce an ABC pattern by clapping, slapping thighs, and making a pop sound. Repeat the pattern so the children can see and hear it.

5. Make several copies of the Snake Pattern (page 79). Glue them onto different colors of construction paper and laminate them. Demonstrate how to use different sorting items found in the Math Center to make colorful patterns on the snakes.

6. Add a poster of patterns to the Math Center. Demonstrate for the children how to read the patterns from left to right using the properties of the patterns (e.g., color) or the ABC language.

7. Use the patterns on page 80 to create a colorful ear of Indian corn. (*Note:* Enlarge the pattern to meet the children's coloring skill levels.) Use the grid paper for the children to color in a pattern of their choice for the ear of corn. Make tracers (stencils) of the corn husk shape and let the children trace and cut out two husks on yellow or manila paper. Cut out the ear of the corn shape and attach the husks on both sides. Display the finished fall decoration.

8. Make a large classroom quilt to display individual patterns. Each child will make a "tie-dye" square for the quilt. Demonstrate how to fold a paper towel in half. Turn the towel one-quarter rotation and fold it in half again. Turn the towel one rotation and fold again. Finally, fold the towel again. (There should be a total of four folds.) The towel should now be a small square shape. Provide three different colors of food coloring (dye) in shallow bowls. Show children how to carefully dip each corner of the folded towel into a different color of "dye." Let the dye soak up a part of the towel. Carefully open the towel to observe a repeating pattern. When the towels have dried, tape them to a large sheet of bulletin board paper to simulate a quilt and display for all to enjoy.

Patterns *(cont.)*

Classroom Activities *(cont.)*

9. At snack time, encourage the students to make a pattern with two or more snack items. Let them tell you about the pattern using ABC language.

10. Send home a copy of Cracker Math (page 81) with each student and a snack-sized bag with 12–15 crackers in three different colors. As each child brings back a finished worksheet, encourage him or her to explain the pattern work to the class during circle time. Display the results for all to see as they come back. (You may want to make this a month-long class project and only send home a few of these at a time.)

11. Older children can participate in a home/school activity. Send home a letter asking each child to bring an item to school which shows a pattern. (See page 78.) At circle time let each child describe the pattern on his or her item using descriptive words and ABC pattern language. (*Note:* It is helpful to spread this activity throughout the month so the children can get excited each day to look at one or two new patterns.)

12. Teach a song that has a repetitive pattern. (*Example:* "Head, Shoulders, Knees, and Toes" or "Old McDonald Had a Farm.") Make a poster of the words to the song so the children can see the patterns that the words make.

Listening Activities

1. Use small colored blocks or similar manipulative items and start a pattern. Ask each child, "What kind of pattern am I making? What comes next?" Encourage the children to answer in full sentences. Start with an AB pattern and proceed to more difficult patterns throughout the year as students grasp the concept. Initially, it is helpful to say the prompt words such as *green, yellow, green, yellow.* As the children get comfortable with patterns, do not say the prompt words.

2. Use body motions to create a pattern. Ask the children to repeat it. Have each child identify the kind of pattern he or she is repeating.

Selected Literature

Beep Beep, Vroom Vroom! by Stuart Murphy (HarperCollins, 2000)

Lots and Lots of Zebra Stripes by Stephen Swinburne (Boyds Mills Press, 2002)

Pattern Bugs by Trudy Harris (Millbrook Press, 2001)

Patterns by Karen Bryant-Mole (Gareth Stevens Publishing, 2000)

Patterns by Peter Patilla (Heinemann Educational Books, 2000)

Patterns by Sara Pistoia (Child's World, 2002)

The Pumpkin Blanket by Deborah Zagwyn (Tricycle Press, 1997)

The Quilt Story by Tony Johnston (Putnam Juvenile, 1985)

Sam Johnson and the Blue Ribbon Quilt by Lisa Ernst (HarperTrophy, 1992)

Captain　　　　　　　　　　　　　　　　　　　**Circle Time Game**

Developing Skill: The players will initiate and/or copy different body motions to create patterns of four repetitions. They will identify who is leading the changes, or who is the "Captain."

Materials

- none

Playing the Game

1. Have the players sit in a circle on the carpet.

2. Select one player to leave the circle and hide his or her eyes.

3. Select another player to be the "Captain." This player must begin a series of different motor activities that are repeated in groups of four. For example, clap, clap, clap, clap (hands); tap, tap, tap, tap (foot); rub, rub, rub, rub (stomach); tap, tap, tap, tap (foot); etc.

4. The other players will copy the "Captain" and when the motion is changed, they must do whatever the "Captain" is doing.

5. The player who was hiding his or her eyes will join the circle and will have to figure out who is leading the patterned activity or who is the "Captain." The motor activities will continue until the right selection has been made.

Hello,

This month we have been learning about patterns. We hope that the children see the patterns that can be found all around them. Please help your child find an item from your home that demonstrates a simple pattern. Have your child bring it to school on _____. Some suggestions are shirts, socks, paper towels, napkins, wrapping paper, and linens.

As each child shares his or her discovery with the class, he or she will "name" the pattern. We have been learning two ways to name patterns. The first is to say the descriptive words (e.g., blue, red, blue, red). The second way is to use ABC pattern language. For example, a red/blue pattern would be an example of an AB pattern. A small, medium, large, small, medium, large pattern would be an example of an ABC pattern. It would be helpful if your child practiced "naming" the pattern at home before sharing it at school with the class.

Have fun with this activity, and thank you for helping your child learn about patterns.

Sincerely,

_____

Snake Pattern

Look at my pattern!

Corn Cob Patterns

Use with Classroom Activity #10 on page 77.

Cracker Math

1. Can you estimate how many crackers are in your bag?

2. Now, count your crackers to find out the *exact* number that is in the bag.

3. Make an AB pattern using two colors of the crackers. Color the AB pattern you made on the squares below.

4. Make an ABC pattern using three colors of crackers. Color the ABC pattern you made on the squares below.

5. Can you make up a different pattern? Color your pattern on the squares below.

6. Now enjoy your crackers. Munch, munch, munch!

7. Bring this paper back to school to share your patterns with your classmates.

Graphing

Educational Objectives: The child will be able to interpret information presented in the visual summary form of a picture or bar graph. The child will be able to read the graph to analyze equal, more, and less representations.

Vocabulary: *Graph*—a diagram used to represent relationships between information

Equal—of the same quantity, size, number, value, degree, intensity, etc.

More—used to compare; meaning greater in number or amount

Less—used to compare; meaning not as much, smaller, fewer

Classroom Activities

1. Introduce the concept of graphing at circle time. Start with an easy graph with two options. Use a large sheet of bulletin board paper. For example, use the title, "Boys and Girls in Our Class." Copy the Boy and Girl Patterns (page 84) so that there will be one pattern for each child in the class. On one side of the graph, attach a boy pattern and write the word *Boys*, and on the other half, attach a girl pattern and write the word *Girls*. Tell the children that the graph will be used find out if there are more boys or girls in the class. Let each child come up, one at a time, and select the correct pattern to attach to the graph with tape. When the graph is complete, ask the children questions about the information the graph provides. Point out that specific information regarding how many boys and girls are in the class can be obtained by counting the patterns. Count the patterns together with the children.

2. Introduce a copy of the Three-part Graph (page 85) that can be used to graph one set of sorting items in the Math Center. Model how to use the graph appropriately. Encourage the children to graph one of the sorting sets and report their findings to the teacher. (*Note:* Choose a set of items to sort, with just three attributes to graph.) If the children are able to do a three-part graph, then gradually add the Four- and Five-part Graphs (pages 86–87).

3. Make a floor graph. Use a large, solid-color vinyl tablecloth. Use a permanent marker to draw a bar graph with space for four or five categories onto the vinyl. Make each box approximately 12 square inches (30 sq. cm) in size. This will be large enough for each child to stand on one square. Present a topic for the class members to graph. (*Examples:* eye color, shoe types, favorite ice cream flavor, etc.) Label the floor graph by placing pictures at the bottom of each column. Make the pictures about the same size as the individual squares on the graph so they can be seen easily. Prepare them before class. One at a time, call each child up to find his or her appropriate spot on the graph. Soon the entire class will be standing on the graph. Ask the children to keep their feet still on the graph as they look around them to interpret the information that is represented on the graph. Discuss the results with the students. (*Notes:* This activity may have to be done twice so every child has a turn standing on the large graph.)

4. Introduce the children to different kinds of graphs throughout the year. Ideas to graph might include the children's favorite flavors of ice cream, book characters, seasons, or cartoon shows. Let each child take part in the graphing by coloring in blocks on the graph, gluing on a picture, adding a sticker, standing on the floor graph, etc. Children will enjoy graphing activities more if they actively participate in filling in the graph and counting the results. Ask the children questions such as, "What did we have the most of? Did we have a tie?"

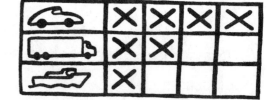

Graphing *(cont.)*

Classroom Activities *(cont.)*

5. You can do graphing throughout the year by keeping several copies of the graphing mats (pages 85–87) printed and available with the teacher's supplies. Use them for the spontaneous teaching moments that happen in the classroom. For example, "What is your favorite way to eat pumpkin?" The headings could include Pumpkin Cake, Pumpkin Cookies, Pumpkin Pie, Roasted Pumpkin Seeds, or Do Not Like Pumpkin. (This is a good time to teach older children the concept of initials. Let each child put his or her initials in the appropriate box on the graph and display for the parents to see at the end of the day. The parents will have fun finding their child's answer on the graph.)

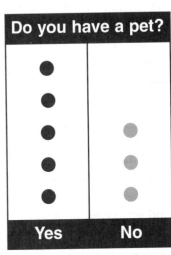

6. After Halloween send a note home asking every student to bring in the wrapping from one of his or her favorite candy treats. Make a graph representing the different kinds of wrapping. For example, some come in boxes, some in little bags, and some in individual wrappers. Make a large graph and let each child glue the wrapper brought to school in the appropriate spot. Display the graph for all to see.

7. Provide a trail mix snack. Make copies of the appropriate graphing mat (pages 85–87) and let the students graph the snack to see which food item they had the most of, least of, etc.

Listening Activities

1. Explain to the children that they will make a three-dimensional graph to find out some new information about the children in the class by using two different colors of snap beads or a similar connecting toy. Present a question with a yes/no option such as "Do you have a family pet?" One color will represent "Yes" and one color will represent "No." Place the words next to the beads. Let the children give their answers and come up one at a time and put the appropriate color of bead on the "Yes" or "No" graph to indicate the choice they have made. After every child has had a turn to contribute to the graph, hold the two structures up for the class to look at to decide how to interpret the information they have organized.

2. Use the large tablecloth graph (see Classroom Activity #3 on page 82). Choose a subject to graph. After the children are all standing on the graph, ask each child one question about the information obtained from the graph. For example, "What does the information on the graph tell us about our friend Abbey?"

Selected Literature

The Best Vacation Ever by Stuart Murphy (HarperTrophy, 1997)

Graph It by Lisa Trumbauer (Yellow Umbrella Books, 2002)

Lemonade for Sale by Stuart Murphy (HarperTrophy, 1998)

Making Graphs by Michelle Nechaev (Gareth Stevens Publishing, 2004)

Treasure Boxes by Jaine Kopp (LHS Gems, 2002)

Boy and Girl Patterns

84

Use with Classroom Activity #2 on page 82 and #5 and #7 on page 83.

Let's graph

Use with Classroom Activity #2 on page 82 and #5 and #7 on page 83.

Let's graph

86

Graphing

Use with Classroom Activity #2 on page 82 and #5 and #7 on page 83.

Let's graph

Parts of a Whole, What's Missing?

Educational Objectives: The child will be able to understand that whole things are made up of many smaller parts. He or she should be able to look at a picture or situation and verbally express the part that is missing.

Vocabulary: *Whole*—containing all of the parts, complete
Missing—something that is absent or lost

Classroom Activities

1. Plan an art activity that includes facial parts. (*Examples:* Jack-o-lanterns, scarecrows, pilgrims, snowmen, Santa Claus, Valentine people, leprechauns, etc.) Demonstrate how to complete the project. As you work, leave off a facial part and ask the children, "What is missing?" Put a facial part on the wrong part of the face and ask, "What is wrong?" Complete the project with the children in small groups. If the children put things in the wrong places, repeat these and similar questions to help them become aware of the parts that make up the whole face.

2. Share the book, *Look Book* by Tana Hoban with the class. This book has die-cut holes in the middle of the page. Show each page to the children and see if they can identify what the picture is, even though they can only see part of it. Lead a discussion about how whole objects are made up of several smaller parts. Then, reveal the whole picture and talk about the small part that was showing through the die-cut hole. Discuss that when only part of an item is showing, it is hard to identify it.

3. Make a set of What's Missing? Cards (pages 89–95). Glue the cards onto construction paper and laminate them for durability. Show the children one card at a time, and have them identify what is missing. Talk about how the missing part would affect the use of the item. Only use a few of the cards; save the rest for the listening activity so each child can have a turn describing what is missing from a card. Place the cards in the Math Center and encourage the children to work with a friend and take turns finding the missing part of the picture.

Listening Activities

1. Use the books by Tana Hoban listed at the bottom of this page and in the "Oldies but Goodies" section (pages 235–236). Show each child one page and let him or her try to identify the picture with part of it hidden. If the child cannot, then show the entire picture and give the child another chance to identify the picture. Encourage each child to use complete sentences when answering. Reemphasize that objects are not easy to identify when parts are missing.

2. Use the What's Missing? Cards (pages 89–95). Show one picture to each child and ask him or her to tell you what is missing and how it would affect the item. Encourage the children to use complete sentences when answering.

Selected Literature

Animal Patterns by Cynthia Cappetta (Innovative Kids, 2001)

Guess What I Am by Anni Axworthy (Candlewick Press, 1998)

Guess Whose Shadow? by Stephen Swinburne (Boyds Mills Press, 1999)

Look Book by Tana Hoban (Greenwillow Books, 1997)

Purple Is Part of a Rainbow by Carolyn Kowalczyk (Children's Press, 1985)

Seven Blind Mice by Ed Young (Puffin Books, 2002)

What Am I? Big, Rough, and Wrinkly by Moira Butterfield (Steck-Vaughn, 1998)

Who's That Scratching at My Door? by Amanda Leslie (Handprint Books, 2001)

88

Use with Classroom Activity #3 and Listening Activity #2 on page 88.

What's Missing? Cards

Parts of a Whole, What's Missing?

Use with Classroom Activity #3 and Listening Activity #2 on page 88.

What's Missing? Cards *(cont.)*

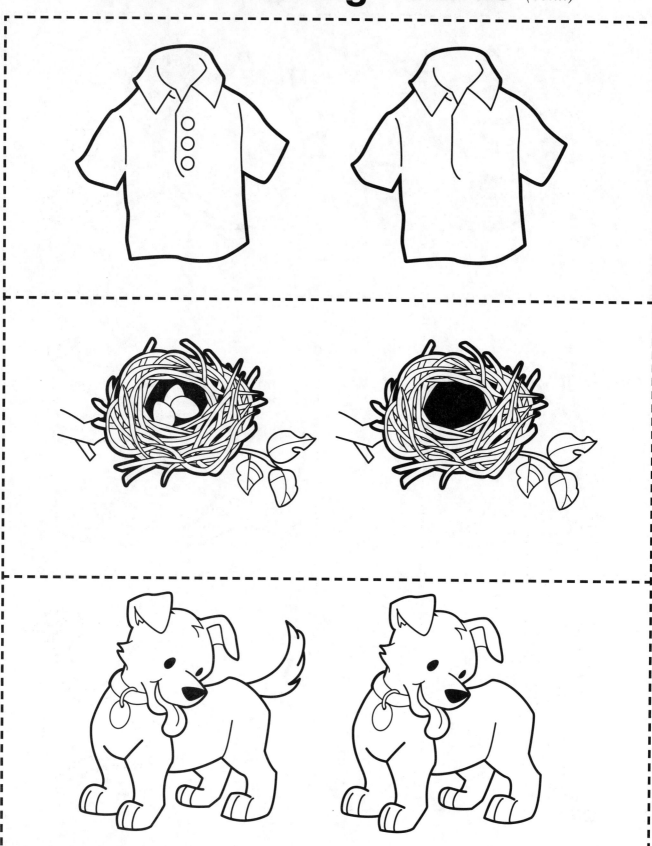

Use with Classroom Activity #3 and Listening Activity #2 on page 88.

What's Missing? Cards *(cont.)*

What's Missing? Cards *(cont.)*

92

Use with Classroom Activity #3 and Listening Activity #2 on page 88.

What's Missing? Cards (cont.)

What's Missing? Cards *(cont.)*

Use with Classroom Activity #3 and Listening Activity #2 on page 88.

What's Missing? Cards *(cont.)*

Ordinal Numbers

Educational Objectives: The child will be able to indicate the place order of an object or event—first, second, third, etc. He or she will be able to use the ordinal numbers *first* through *tenth*.

Vocabulary: *Ordinal number*—any number used to indicate place order in a particular series

Classroom Activities

1. Make a copy of the Ordinal Numbers Poster (pages 98–101). Color the children's clothes and accessories with bright colors and glue the pages together to make one long poster. Mount the paper on construction paper and laminate it. Use the poster to teach the ordinal numbers *first* through *tenth*. Attach the poster to the wall in the Math Center at a low level so the children will be able to touch the numbers to review them.

2. Call three or four of the children to the front of the circle and label who is first, second, third, etc. Do this several times so all the children will have a special turn to be called on. This will reinforce the new vocabulary that you are using.

3. Provide a snack that the children can put together before they eat it. Hand out the directions and then demonstrate how the children follow a specific order to put the snack together. *Example:* Frosty Fruit (see page 102)—First, crumble a graham cracker into a bowl. Second, cover the crackers with a spoonful of whipped cream or yogurt mixed with fruit cocktail. Third, sprinkle with miniature marshmallows. Fourth, eat and enjoy the snack!

Listening Activities

1. Let each child have a turn playing Carnival Cups (see page 97).

2. Arrange several plastic animals, dolls, bugs, etc., in a long line. Identify which toy is the first in the line. Ask each child to identify the first toy, second toy, third toy, fourth toy, etc., Encourage the children to use complete sentences when describing the toy.

3. Use the Ordinal Numbers Poster (pages 98–101) and ask each child a question using the poster:

   - Which child is holding four balloons?
   - Which child is walking a dog?
   - Which child is holding a soccer ball?
   - Which child is carrying a suitcase?
   - Which child is using inline skates?
   - Which child is pulling a toy duck?
   - Which child is pushing a shopping cart?
   - Which child has a monkey on his shoulder?
   - Which child is holding a turtle?
   - Which child is behind a wall?

   This activity can also be reversed and the teacher can ask, "What is the third child doing?"

4. Ask each child about the position of different letters in his or her name. (*Example:* "Tell me the third letter in your name.") Provide a dry-erase board so if the child cannot do this independently, he or she can write the name on the board and figure out the answer.

Selected Literature

Get Well, Good Knight by Shelly Moore Thomas (Puffin Books, 2004)

Henry the Fourth by Stuart Murphy (HarperTrophy, 1999)

Seven Blind Mice by Ed Young (Philomel Books, 1992)

The Twelve Days of Kindergarten by Deborah Rose (Harry N. Abrams, 2003)

Carnival Cups   2 players

Developing Skill: Players will use ordinal numbers to identify the cup they think has the hidden item.

Materials

- 4 plastic, solid-colored cups
- 1 small item that can be covered by a cup (pompom, small eraser, etc.)
- 1 table or small tray (to position the cups on so they can be easily moved)

Playing the Game

1. One player places the chosen item under the first cup. The other player watches closely as the first player slides all four cups around on the tabletop so they are lined up in different positions.

2. The second player tries to guess which position the hidden object is in using the words *first, second, third,* or *fourth.* If the answer is not correct, he or she continues guessing until the object is found.

3. If the answer is correct, the players exchange positions and play continues back and forth.

Ladybugs in the Lighthouse   2–4 players

Developing Skill: Players will move a ladybug from the first to the fifth level of the lighthouse, identifying each level using ordinal numbers.

Materials

- Lighthouse pattern (pages 103–104)
- 4 different-colored ladybugs (plastic, rubber, wood, etc.)
- 1 standard die (dots 1–6)

Preparation: Color the light on the pattern a bright yellow. Highlight the five different levels on the lighthouse with yellow as well. Attach the pages together to create a tall lighthouse. Glue the lighthouse onto construction paper or tagboard and laminate it for durability. If the ladybugs are not different colors, add some color so they can be identified as different markers.

Playing the Game

1. Each player places his or her ladybug on the start space at the bottom of the lighthouse. Select who will go first and the order of play.

2. Each player rolls the die to see how many spaces he or she will move up through the different levels to get to the bright light.

3. As the players advance up the levels, they should announce which level they are on. For example, "I'm on the *second* level!"

4. Play continues until all the bugs are on the light and it is time to "party." (**Note:** When modeling how to play this game, use the ordinal numbers vocabulary.)

Use with Classroom Activity #1 on page 96.

Ordinal Numbers Poster

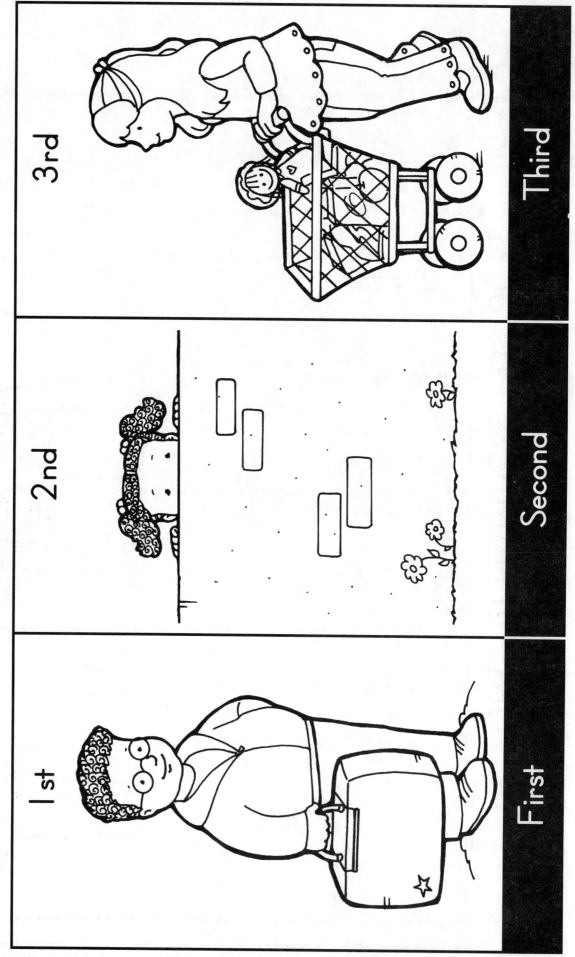

Use with Classroom Activity #1 on page 96.

Ordinal Numbers Poster *(cont.)*

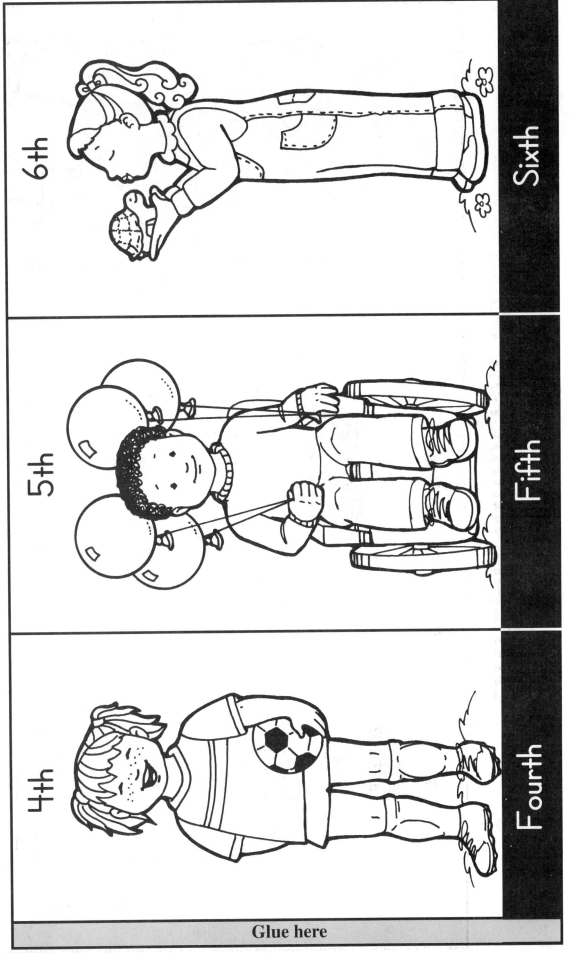

Use with Classroom Activity #1 on page 96.

Ordinal Numbers Poster *(cont.)*

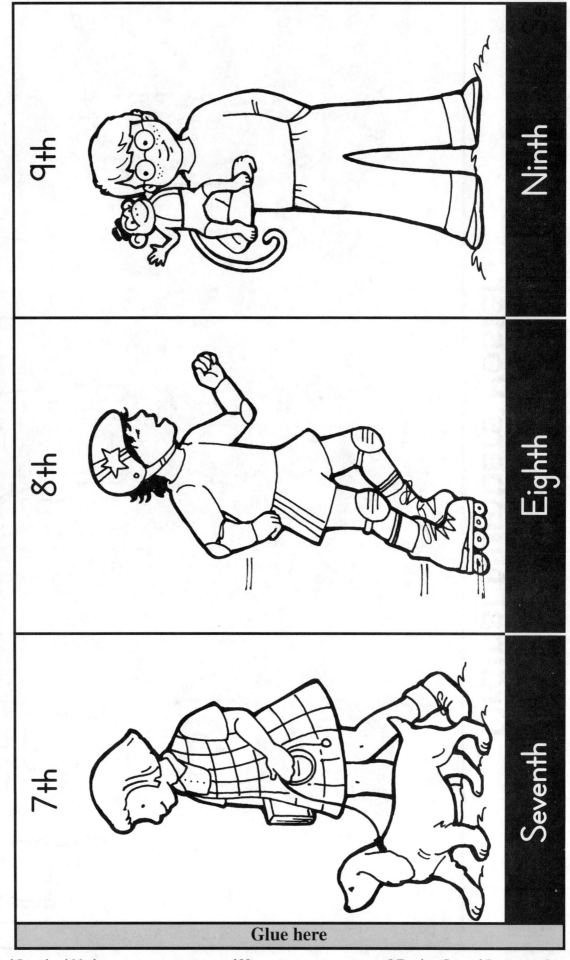

Glue here

Use with Classroom Activity #1 on page 96.

Ordinal Numbers Poster *(cont.)*

Numbers

Ordinal

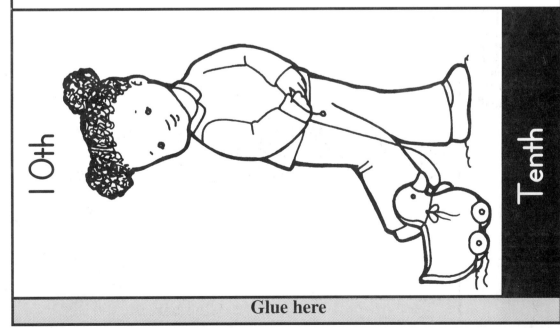

10th

Tenth

Glue here

Frosty Fruit

Crumble a graham cracker into the bowl.

1st

Spoon fruit mixture onto the crackers.

2nd

Sprinkle with miniature marshmallows.

3rd

Eat and enjoy the snack.

4th

Lighthouse Pattern: Use with the Ladybugs in the Lighthouse game on page 97.

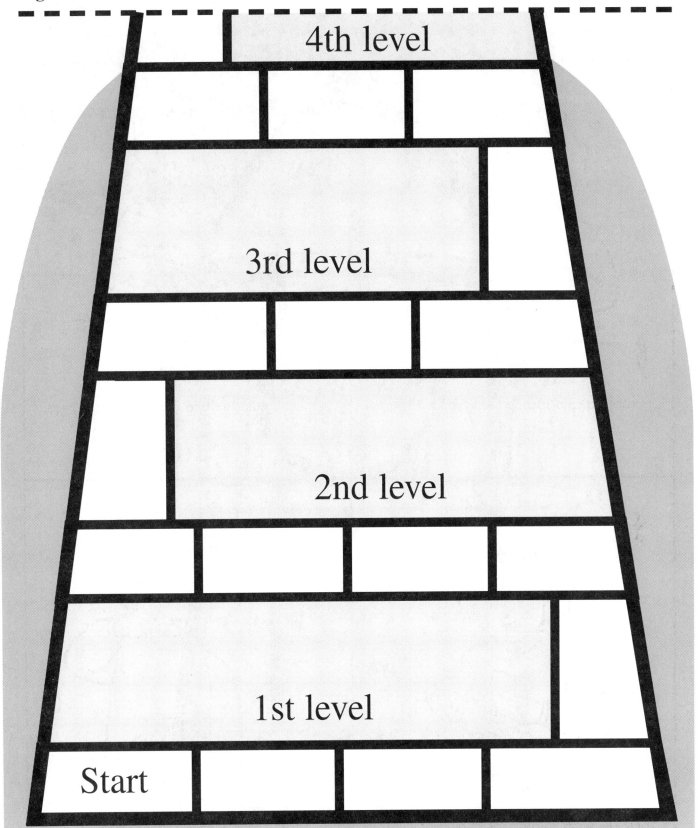

4th level

3rd level

2nd level

1st level

Start

Ladybugs in the Lighthouse

Lighthouse Pattern: Use with the Ladybugs in the Lighthouse game on page 97.

Sequencing

Educational Objectives: The child will be able to place three or four events or pictures in a logical order and be able to explain his or her actions.

Vocabulary: *Sequence*—the following of one thing after another in a logical order

Classroom Activities

1. Introduce the concept of sequencing with the visual aid of sequence cards. Use the cards to illustrate how an event must be completed in a logical order to make sense. Use the words *first, second, third,* and *fourth* (review the Ordinal Numbers section on pages 96–104) to help place the pictures in order. The sequence cards can be commercial or you can use the Sequencing Cards (pages 108–110).

2. Cooking activities are a great way to help the children understand the concept of completing work in a logical order. Point out to the children that each step needs to be completed in the proper sequence for the food to be prepared properly. Plan to prepare a food item with the students. Pizza, fruit salad, spaghetti, tacos, pancakes, applesauce, and vegetable soup are all good examples. Let the children make their own recipe book with step-by-step directions. *Example:* Make biscuit pizzas using refrigerator biscuits (see pages 111–112).

3. Older children will enjoy completing a project on their own by following a sequence chart. Introduce a simple art activity that the children will be able to complete independently (see page 113). Show the children the sequence of the directions at the top of the page. Have the children follow as the teacher demonstrates each step of the project. Explain that the children will put together a puzzle independently by doing the steps in the proper sequence.

4. Demonstrate a cut-and-paste sequence activity at circle time (see pages 108–110). Provide construction paper that will be long enough for the children to paste the pictures in the correct sequence, placing them from left to right. After the pictures have been put in the proper sequence, have each child tell the story to the teacher. Begin with three-part events, then move up to four- and six-part sequence activities when the children are ready for more challenging work.

5. Sequencing can be incorporated into many science units. Make a mobile showing the four stages of a butterfly's life (page 114), hanging from top to bottom. Discuss other animals that develop in a special sequence (e.g., frogs, bees, etc.). Another science idea would be to make a poster showing the stages of a plant's growth, starting with a seed and sequencing from left to right to a mature plant (see page 115).

6. Let the children decide how they want to sequence a story of their own. It is fun to do this with story symbols. Demonstrate how to do this activity at circle time. Use a flannelboard to put up story symbol pictures that can be used to tell a simple story in a logical sequence. *Example:* One morning a girl woke up, ate her breakfast, brushed her teeth, and combed her hair to get ready for school. Then give each child a 12" x 24" piece of construction paper (30 cm x 61 cm). Provide an assortment of story symbols (see pages 116–117). These can be individual pictures or stencils to trace or use with paint. Encourage each child to choose symbols and place them in a sequence from left to right to tell a story. When the stories are complete, have the children assemble in a circle and let everyone have a chance to tell his or her story.

7. Add commercial or homemade sequencing cards to the Math Center.

Sequencing *(cont.)*

Listening Activities

1. Give each child four sequence cards to place in the proper sequence. Ask the child to tell the story that the cards tell to his or her classmates. (You can use commercial sequence cards or homemade ones.)

2. Describe for each child a list of three actions. Ask the child to do them in the sequence that they were stated. *Example:* Stand up, go to the chalkboard, and put your hand on the chalkboard. (It will help to have a list prepared beforehand so you don't have to think quickly.) The children are usually excited to have their turn with this silly activity.

Selected Literature

The Giant Jam Sandwich by John Lord (Houghton Mifflin, 1991)

The Gingerbread Baby by Jan Brett (Putnam Publishing Group, 1999)

The Gingerbread Boy by Richard Egielski (HarperCollins, 2001)

The Hungry Farmer by Michelle Nechaev (Sagebrush, 2003)

I Know an Old Lady Who Swallowed a Pie by Alison Jackson (Puffin Books, 2002)

If You Give a Moose a Muffin by Laura Numeroff (HarperCollins Children's Books, 1991)

If You Give a Mouse a Cookie by Laura Numeroff (HarperCollins Children's Books, 1985)

If You Give a Pig a Pancake by Laura Numeroff (HarperCollins Children's Books, 1998)

If You Take a Mouse to the Movies by Laura Numeroff (HarperCollins Children's Books, 2000)

If You Take a Mouse to School by Laura Numeroff (HarperCollins Children's Books, 2002)

There Was a Cold Lady Who Swallowed Some Snow by Lucille Colandro (Scholastic, Inc., 2003)

The Tiny Seed by Eric Carle (Simon & Schuster Children's Books, 1991)

The Tortilla Factory by Gary Paulsen (Harcourt, 1998)

The Very Busy Spider by Eric Carle (Philomel, 1995)

The Very Hungry Caterpillar by Eric Carle (Philomel, 2001)

The Wolf's Chicken Stew by Keiko Kasza (Grosset & Dunlap, 1996)

Ice-Cream Cone Game

2–4 players

Developing Skill: Players will need to understand the sequence of a melting ice-cream cone so they can arrange their cards in the proper order.

Materials
- Ice-Cream Cone Game Cards (pages 118–119)

Preparation: Make 4 copies of the game cards. Color the cones brown. Next color each set of scoops with a different flavor of ice cream. Cut the six cards apart, glue them onto construction paper, and laminate them for durability. Color four copies of the completed sequences to match the four flavors of ice cream. These will be used as references for the children. Glue them to construction paper and laminate them for durability.

Playing the Game

1. Each player will need to complete a six-part sequence showing the melting of an ice-cream cone. To begin, each player selects a full cone of a different flavor. This will determine what color (flavor) the rest of his or her playing pieces will be.

2. All of the ice-cream cone cards are shuffled and put in a pile.

3. The first player picks the top card. He or she is looking for the second picture in the sequence to get started. If the card selected is not the one he or she needs, it is placed faceup in a discard pile.

4. The next player takes a turn. At each turn, the player can select from the top of the discard pile or get a new card from the top of the original pile. The discard pile will need to be turned over to go through a second or third time until all the ice-cream cone sequences have been completed.

5. Play continues until everyone has completed the melting-cone sequences.

Santa's Elf

Circle Time Game

Developing Skill: Players must remember the sequence of toys that have been made by each of their classmates.

Materials
- none

Playing the Game

1. Have the players sit in a circle on the carpet. This is a memory sequence game.

2. The first player says, "I'm Santa's Elf and I make_____." The player says a kind of toy and makes a hand motion that would be associated with the toy. *Examples:* doll (rocking baby motion), train (pulling a whistle cord, choo, choo), race car (steering), basketball (dribbling), etc.

3. The second player says what he or she is making, and then repeats what the previous player said. The third player says a new item and repeats what the other two before him or her said in the correct sequence.

4. Continue around the circle until everyone has had a turn. The last player has to remember what everyone has already said. The hand motions make this a visual, as well as an auditory game.

Use with Classroom Activities #1 and #4 on page 105.

Three-part Sequencing Cards

Use with Classroom Activities #1 and #4 on page 105.

Four-part Sequencing Cards

Six-part Sequencing Cards

Use with Classroom Activity #2 on page 105.

Recipe Book Cover

Pizza Recipe Book

Pizza Recipe Book

Pizza Recipe Book

Pizza Recipe Book

Pizza Recipe Book

Pizza Recipe Book

Pizza Recipe

Put biscuit on wax paper.
Flatten and spread the dough.

Spread pizza sauce on the biscuit.

Put three (3) pepperoni slices on top of the sauce.

Sprinkle cheese on top of the pepperoni.

Cook in the oven 350ºF for 15 minutes.

Eat your pizza. Yum!

Sequencing

Use with Classroom Activity #3 on page 105.

Pumpkin Puzzle

and Paste and =

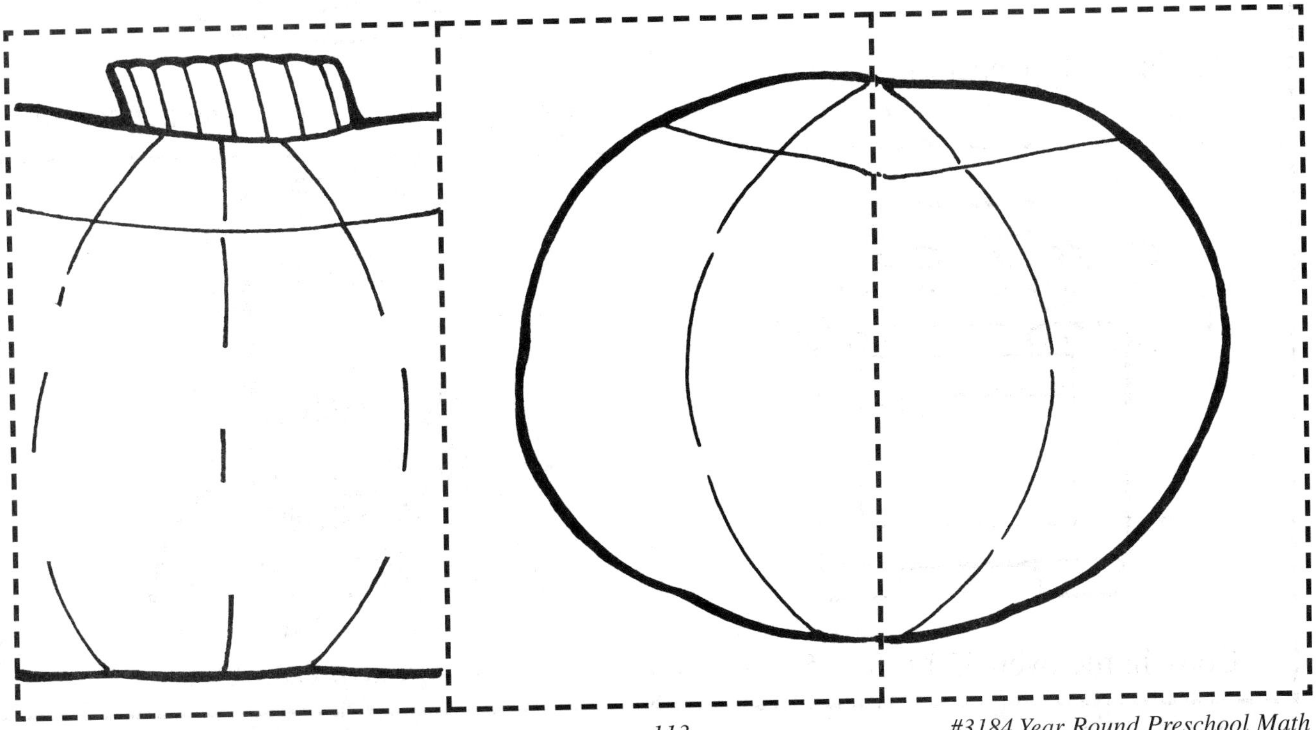

©Teacher Created Resources, Inc.

113

#3184 Year Round Preschool Math

Butterfly Sequencing

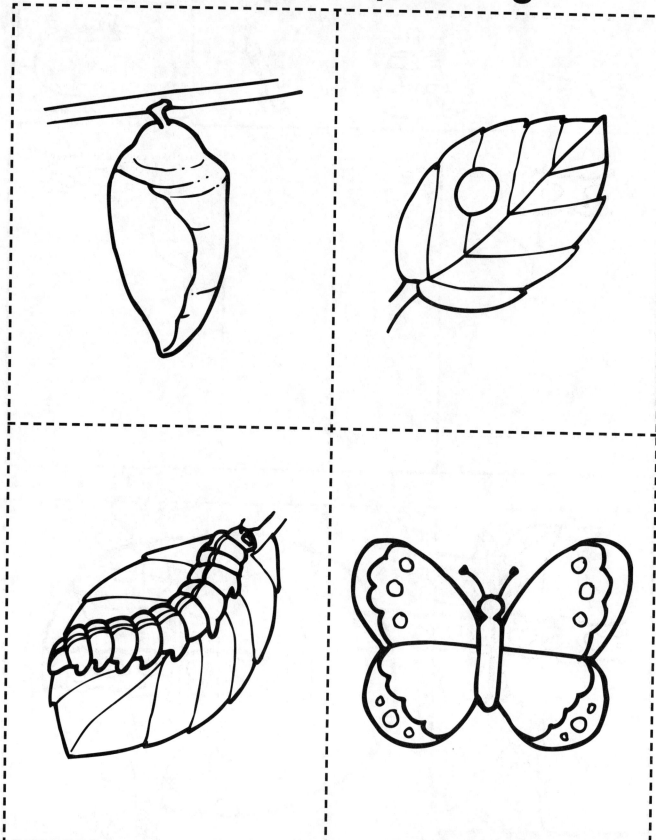

114

Use with Classroom Activity #5 on page 105.

Growing Seed Sequencing

Story Symbols

Use with Classroom Activity #6 on page 105.

Story Symbols *(cont.)*

Ice-Cream Cone Game Cards

Use with the Ice-Cream Cone Game on page 107.

Ice-Cream Cone Game Cards *(cont.)*

Seriation

Educational Objectives: The child will be able to sort and arrange objects according to "how much" of the property the object has (big to small, fat to thin, heavy to light, etc.).

Vocabulary: *Seriation*—an established order of how much of a property the object has

Classroom Activities

1. Introduce the concept of seriation with a familiar item in three different sizes. Demonstrate the proper way to seriate using the words *small, medium,* and *large,* arranging the items from left to right. (Toddler nesting cups are a good visual to use to introduce this concept.). Introduce another nesting cup and demonstrate where it would fit in the grouping. Is it a little smaller or bigger? Work up to five or six items. Demonstrate how to line up all the items on a common line to help see the difference in sizes. (Present this activity often throughout the school year with different examples of the concept of seriation such as tallest to smallest, longest to shortest, heaviest to lightest, etc.)

2. Place a container of various items to seriate in the Math Center. (Example: measuring cups, measuring spoons, nesting dolls, dowels cut to different lengths, tapered candles burned down to different sizes, straws of different lengths, etc.)

3. Line the class up from tallest to shortest. Remember to include the teacher. (Do not do this activity if you have one child who is exceptionally short or tall, as it could become an uncomfortable experience.)

4. Call four girls with different lengths of hair up to the front. Ask the class to place them in order from shortest to longest hair.

5. Bring a building toy to the circle and demonstrate how to build things of different lengths from shortest to longest, etc. This will encourage children to continue working on this skill on their own during free play.

6. Seriated Evergreen Tree 1—Start with a 9" x 12" (23 cm x 30 cm) piece of green construction paper. Fold one corner to the spot on the rectangle that will create a triangle. Cut off the excess so when you open the paper, you end up with a square shape. (Keep the cut off part to use later for the tree trunk.) Cut on the fold mark to create two matching triangles. Set one triangle aside and fold the other triangle in half again and cut on the fold. Once again, set one triangle aside and fold the other triangle and cut. Repeat until the triangle is too small to work with. Use the last triangle for the top of the tree and discard the other tiny piece. Arrange the triangles from largest to smallest on another piece of paper to create an evergreen tree shape. Secure the pieces with glue. Cut out a trunk using the first discarded piece of paper and add it to the bottom of the tree. This tree can be colored and decorated in many different ways.

Seriation *(cont.)*

Classroom Activities *(cont.)*

7. Send home a letter (page 122) describing a homework activity called "Fabulous Family Footprints." When the papers come back to school, display them for all to admire.

8. Seriated Evergreen Tree 2—Make a pattern of an isosceles triangle that will fit onto a 9" x 12" (23 cm x 30 cm) piece of green construction paper with the base of the triangle on the short side of the paper. Trace and cut out the tree shape. Fold the tree in half by bringing the point of the tree down to the bottom of the tree. Fold in half again by bringing the top to the bottom again. Repeat. Open the folded tree and cut on all of the fold lines. Arrange the cut pieces on another piece of construction paper to recreate the evergreen tree shape. Shapes should go from longest to shortest. (Younger children can enjoy this activity by folding only two times.)

Listening Activities

1. Copy the Turkey Patterns (page 123) several times. Laminate them for durability. Put three to five of the turkeys on the floor in front of a child and ask him or her to seriate them from largest to smallest or smallest to largest. Use a different selection of turkeys for each child. Provide a paper with a line on it to assist the children. Have the children arrange the turkeys from left to right. Ask each child to explain to the class how he or she chose to seriate the turkeys.

2. Use an assortment of the materials from the Math Center. Let each child select which item he or she would like to seriate and the order. Have the child explain his or her work when it is completed.

Selected Literature

The Crunching Munching Caterpillar by Sheridan Cain (Tiger Tales, 2003)

The Doorbell Rang by Pat Hutchins (HarperTrophy, 1989)

Mr. Willowby's Christmas Tree by Robert Barry (Random House Children's Publishing, 2000)

One. . . Two. . . Three. . . Sassafras! by Stuart Murphy (HarperTrophy, 2002)

There Was an Old Lady Who Swallowed a Fly by Simms Tabeck (Viking Books, 1997)

There Was an Old Lady Who Swallowed a Trout by Teri Sloat (Henry Holt & Company, 1998)

The Three Billy Goats Gruff by Stephen Carpenter (HarperCollins Children's Books, 1998)

Titch by Pat Hutchins (Aladdin, 1993)

Old Lady Who Swallowed a Fly   **2–4 Players**

Developing Skill: Players need to be able to collect the animals the old lady swallowed in the correct order, starting with the smallest and ending up with the largest.

Materials

- Old Lady Who Swallowed a Fly Game Cards (pages 124–125)

Preparation: Make four copies of the game cards. Color in the nine individual playing cards on each sheet. Cut them apart and glue them onto construction paper. Laminate them for durability. Color in the four long sequence pictures using the same color selections. Glue this onto construction paper and laminate. These will be used as guides for each child to follow.

Playing the Game

1. Each player selects an old lady card and a long picture card with the animals seriated in the correct order. These cards are placed on the table in front of the player.

2. Shuffle the playing cards and place them in a stack.

3. Select who will go first. This player takes the top card on the pile. If it is a fly, he or she keeps it, places it next to the lady, and draws again. This time he or she will be looking for a spider. If the first card selected is not a spider, then the player places it faceup in a discard pile.

4. It is the next player's turn. He or she can choose the top card from either pile. Each player is looking for the fly to get started.

5. Play continues until each player has lined up the old lady and all the animals she swallowed in the correct order from the smallest to largest.

Hello,

This month we will be introducing the math skill of *seriation*. This is the ability to arrange items according to their properties such as longest to shortest, heaviest to lightest, smallest to biggest, etc. To help the children grasp this concept, we have developed an activity for you to do with your child at home. We are sending home a colorful piece of paper for you to use to trace one footprint for each family member. Family pets can be included in this project if they choose to participate!

Ask your child if he or she wants to start with the biggest or the smallest family member. Then have him or her direct the order of tracing the feet. Have each family member place the heel of his or her foot on the long side of the paper. Use a dark marker to trace his or her foot. Next, decorate the tracing. It is fun to let each family member add a little personal touch to his or her own footprint. Footprints can be covered with pictures of hobbies, stickers, etc. Label the top of each foot with the name of the family member, and add the family name at the top of the paper. Send the finished project back to school the next time our class meets. We will make a fun display in the hall called "Fabulous Family Footprints." Thank you for all of your assistance.

Sincerely,

_____

Use with Listening Activity #1 on page 121.

Turkey Patterns

Old Lady Who Swallowed a Fly

Use with the Old Lady Who Swallowed a Fly game on page 122.

Old Lady Who Swallowed a Fly *(cont.)*

Time

Educational Objectives: The child will be able to recognize that various devices are used to measure time. He or she will be able to identify several items that measure time, such as digital and analog clocks, a timer, a watch, a minute (sand) timer, and a calendar. The child will be able to identify the hours on a clock and distinguish between activities that happen during the day and the night.

Vocabulary: *Clock*—an instrument used for measuring and indicating time

   Timer—an instrument used for measuring short amounts of time
        (seconds, minutes, hours)

   Watch—a small timepiece (clock) designed to be worn on the wrist

   Calendar—a chart that arranges the year in days, weeks, and months

   Day—the period of time between sunrise and sunset

   Night—the period of time from sunset to sunrise

Classroom Activities

1. Introduce time by showing several items used to tell time. (*Examples:* different kinds of clocks, watch, stopwatch, sand timer, baking timer, calendar, sundial, etc.) Explain that the children are going to learn about telling time with a clock. Using a large educational clock, identify the parts of the clock, including the face, the hands, and the numerals. Explain that the clock can tell hours and minutes. Point out that every numeral on the face of the clock represents one hour. Show the small lines that represent the minutes. Move the hour hand on the clock to represent different hours and ask the children what the different times are.

2. Explain to the children that long ago before clocks, people used other tools to help tell time. Show the students a candle clock, place it in a safe place in the classroom, light it, and let the children watch it during their class time. The children will enjoy watching the candle burn down and the pins drop as each hour passes by. To make a candle clock, use a long tapered candle. Put push pins down one side of the candle approximately 1" (2.54 cm) apart. As the candle burns, the pins drop down and make a sound. (Make a candle clock at home first to determine how long it takes for the candle to burn down one hour to make your clock even more realistic.)

3. Help the children understand how long one minute is by demonstrating this One Minute activity to the class. (One- and two-minute timers are available at teacher supply stores.) Use a minute (sand) timer, a plastic set of tweezers, a plastic bowl, and a plastic jar filled halfway with a small item. (Use something soft, such as pompoms, for smaller children who are just learning to use tweezers. Use a hard item, such as buttons, for older children.) Ask the children to estimate how many items they will be able to pick up with the tweezers and put in the bowl before the timer empties (1 minute). Only one item may be picked up at a time. Demonstrate how to do this activity by choosing a child to be the "Timekeeper" while you do the activity. He or she will turn over the timer and say, "Go" and then, "Stop" when the timer is empty. Together count the items removed from the jar. Add this activity to the Math Center. Encourage the children to take turns being the "Timekeeper."

4. As a class, make up a time line of the activities that happen at school, placing them in the proper sequence. Add the hour when each activity happens. Prominently display the time line.

5. Add an educational clock and a clock poster to the Math Center. (A clock poster can be made using some of the cards on pages 129–131.) Hang the poster low enough so the children can touch the clocks and read the times that they represent.

Time *(cont.)*

Classroom Activities (cont.)

6. Use the Clock Concentration Cards (pages 129–130) to practice reading clock faces. Make two sets of the analog and digital clock cards. First, arrange concentration (Memory) games where children find matching digital or analog clocks. Later as they become more adept at reading the clocks, play a game where analog clocks are matched to digital clocks. Younger children can play similar games with the cards arranged face up instead of face down.

Listening Activities

1. Use a large educational clock. Set the minute hand at the 12 and the hour hand on any number. Display a different time for each child and ask him or her, "What time is it?"
2. Display different items that are used to tell time and ask each child to name the item. (*Example:* wristwatch, stopwatch, sand timer, baking timer, calendar, sundial, etc.)
3. Use the Clock Concentration Cards (pages 129–130). Display several clocks on an easel or flannelboard. Call on one child at a time and ask him or her to choose a clock. Ask, "What time is it?'
4. Make up a list of events that are familiar to the children in the class. Ask if each event takes place during the day or the night. (*Example:* I sleep during the . . . , I go to school during the . . . , etc.)

Selected Literature

Bats Around the Clock by Kathi Appelt (HarperCollins, 2000)

Big Hand, Little Hand by Judith Herbst (Barron's Educational, 1997)

Bunny Day by Rick Walton (HarperCollins, 2002)

Clocks and More Clocks by Pat Hutchins (Simon & Schuster, 1994)

Cluck O'Clock by Kes Gray (Holiday House, 2004)

The Grouchy Ladybug by Eric Carle (HarperCollins, 1996)

Monster Math Schooltime by Grace Maccarone (Scholastic, Inc., 1997)

P. Bear's New Year's Eve Party by Paul Lewis (Ten Speed Press, 1999)

Telling Time by Winky Adam (Dover Publications, 2000)

Telling Time with Big Mama Cat by Dan Harper (Harcourt Childrens Books, 1998)

Telling Time with Tickety Tock by Sarah Landy (Spotlight/Nickelodeon, 2001)

Time by Sara Pistoia (Child's World, 2002)

Time by Henry Pluckrose (Children's Press, 2001)

What's the Time? by Lara Tankel Holtz (DK Publishing, 2001)

What Time Is It? by P. D. Eastman (Random House Books for Young Readers, 2002)

What Time Is It? by Sheila Keenan (Scholastic, Inc., 2000)

What Time Is It, Mr. Crocodile? by Judy Sierra (Harcourt, 2004)

The Clock Game   **2–4 Players**

Developing Skill: The player will need to know that each numeral on the face of the clock represents one hour. He or she will need to be able to move the hour hand on the play clock to correspond with the number of hours he or she spins on the spinner.

Materials

• Clock Pattern (page 131)

• 1 spinner with four sections (0–4)

Preparation: Copy the Clock Pattern onto four different colored pieces of construction paper. Cut out the clocks and the hour hands. Glue two hour hands together for each clock. Laminate the clocks and the separate hour hands. Attach the hour hands to the clocks with brads. The spinner can be store-bought or made using a plastic lid from a food item, such as a coffee can or raisin container lid and a brad, paper clip, and small metal washer. Divide the lid into four equal sections. Leave one section empty and place stickers on the other sections to indicate 1, 2, and 3. Poke a hole through the center of the lid. Push the brad through the end of the paper clip, the washer, and the plastic lid. Secure the brad.

Playing the Game

1. All the players start with their clock showing 12:00. The object of the game is to be the first one to move the hour hand all the way around the face of the clock and back to 12:00.
2. The first player spins the spinner to see how many hours he or she will be able to move the clock's hour hand. The player should count aloud and announce what time it is at the end of each turn. (*Example:* "Four o'clock.")
3. Play continues until one player has moved his or her hour hand back to the 12:00 spot.

Clock Concentration   **2–4 Players**

Developing Skill: The player will need to be able to read the time on the analog and digital clocks so he or she can find matching times on the cards. He or she will need to remember where different clocks are placed on the table so a match can be found.

Materials

• Clock Concentration Cards (pages 129–130)

Preparation: Copy the cards onto construction paper. Laminate them for durability. Cut apart the playing cards.

Playing the Game

1. Place 4–12 matching card sets (determined by skill level) facedown in four rows.
2. Decide who will go first. This player turns over two cards and places them on the table so all the players can see them. He or she will try to find the same time on a digital clock and an analog clock. If the cards match, he or she gets to keep them and take another turn.
3. If they do not match, the cards are turned back over and another player takes a turn. As more cards are revealed, it is easier for the players to find matching cards.
4. Play continues until all the cards have been collected. The player with the most cards wins.

Use with Classroom Activity #5 on page 126 and the Clock Concentration game on page 128.

Clock Concentration Cards

Clock Concentration Cards *(cont.)*

Use with Classroom Activity #5 on page 126 and the Clock Concentration game on page 128.

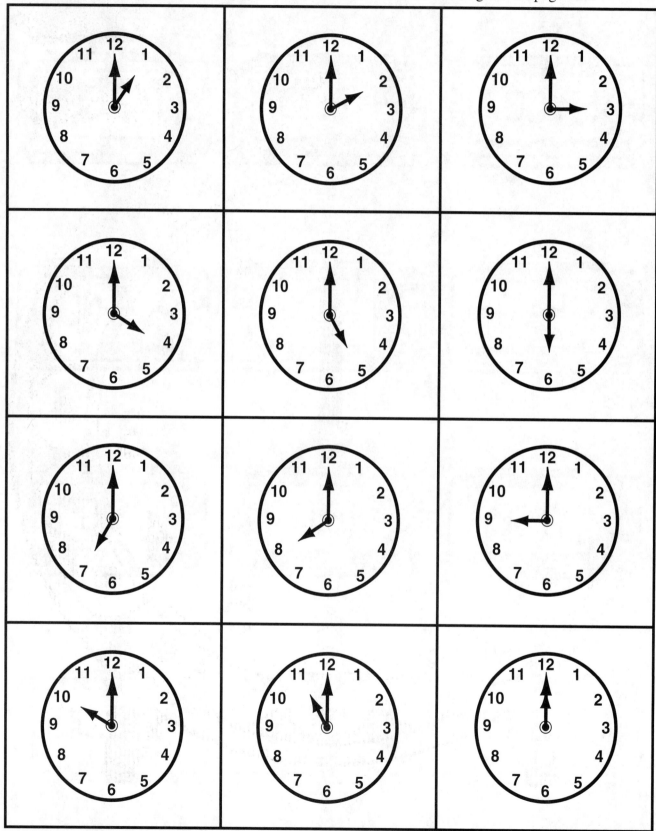

Use with The Clock Game on page 128.

Time

Clock Pattern

Flat Shapes

Educational Objectives: The child will be able to identify the following eight shapes: circle, square, rectangle, triangle, oval, hexagon, rhombus (diamond) and semicircle.

Vocabulary:

● *Circle*—a shape that is round and closed, without sides or corners

■ *Square*—a shape with four corners and four equal sides

▬ *Rectangle*—a shape with four corners and four sides

▲ *Triangle*—a shape with three sides and three corners

● *Oval*—a rounded shape that is longer or wider than a circle

⬡ *Hexagon*—a shape with six sides

◆ *Rhombus (Diamond)*—a tilted shape with four equal sides

⬤ *Semicircle*—a shape that is half of a circle

Classroom Activities

1. Introduce shapes to the class using the large, flat, animated sample of each shape (pages 135–138). Copy each shape on a separate sheet of paper and laminate it. Point out similar shapes in the classroom. (*Examples:* the clock is a circle, the book is a rectangle, etc.) Introduce one shape at a time. The study of shapes can be spread out over the entire year. Review the first four shapes mentioned above before introducing the next four shapes.

2. Create a poster with labeled flat shapes. Place the poster in the Math Center.

3. Make a seasonal flannelboard activity. (*Example:* Cut out a large felt Christmas tree. Use different colors of felt to cut out ornaments of different shapes. Also cut out different shaped presents to place under the tree. Let the children have fun decorating the shape tree.)

4. Add colorful shape design cards for the geoboards in the Math Center. These will encourage the children to copy exact shapes, rather than just make random designs with the rubber bands. These can be store-bought or handmade.

5. Provide stencils of flat shapes in the Art Center. Encourage children to trace them and create their own unique artwork using the shapes.

6. Use the flat shapes on pages 135–138 to make practice sheets for use with play dough. Glue the pages onto construction paper and laminate them. Demonstrate how to roll out snakes with the dough to cover the shape outline.

7. Have a Shape Hunt. Cut out several different colors and sizes of the eight shapes. Hide the shapes while the children are out of the room. When the children return, give them a limited amount of time to find the shapes. Then encourage children to make a collage using the found shapes. (**Hint:** Keep back a few shapes in case a child does not find any. Then, discreetly help this child find them.)

8. Serve snacks that are in the shapes you are studying. (*Example:* Cut cheese into rhombus shapes, serve square crackers and circle pepperoni, etc.)

9. Add chalk and patterns to the Math Center so the children can draw shapes on the chalkboard. If you have a playground, let the children take the chalk outside to draw shapes on the sidewalk. Provide stencils for the children to trace.

Flat Shapes *(cont.)*

Listening Activities

1. Use the eight laminated shapes from classroom activity #1 on page 132. Place the assorted shapes where all of the children can see them. Review classmates' names by asking each child to select a shape and give it to another child. Have the child explain what he or she is doing such as, "I'm giving a square to Beth."

2. Review descriptive words by asking each child to put a shape in a specific place. Use words that refer to position such as *next to, underneath, beside, on top, above, in front of, behind, below, across,* etc. Ask the child to repeat the words as he or she completes the task (i.e., "I put the oval on the table.").

3. When the class has demonstrated a good understanding of the eight shapes, let each child play "teacher." Have each child select a shape and hold it up for a friend to identify (i.e., "What shape is this, Becky?").

Selected Literature

Bear in a Square by Stella Blackstone (Barefoot Books, 2000)

The Greedy Triangle by Marilyn Burns (Scholastic, Inc., 1995)

Ovals by Jennifer Burke (Children's Press, 2000)

Ovals by Mary E. Salzmann (Sand Castle, 2000)

Ovals by Sarah Schuette (Capstone Press, 2002)

Pigs on the Ball by Amy Axelrod (Aladdin, 2000)

Rectangles by Jennifer Burke (Children's Press, 2000)

Rectangles by Sarah Schuette (Capstone Press, 2002)

Round Is a Mooncake by Roseanne Thong (Chronicle Books, 2000)

The Shape of Things by Dayle Dodds (Candlewick Press, 1996)

Shapes by Karen Bryant Mole (Chronicle Books, 2000)

Shapes, Shapes, Shapes by Tana Hoban (HarperTrophy, 1996)

Squares by Jennifer Burke (Children's Press, 2000)

Triangles by Jennifer Burke (Children's Press, 2000)

Twizzlers: Shapes and Patterns by Jerry Pallotta (Scholastic, Inc., 2002)

When a Line Bends . . . a Shape Begins by Rhonda Greene (Houghton Mifflin, 2001)

Tom Turkey Game

2–4 Players

Developing Skill: The players will need to match drawings of the eight flat shapes on the turkey gameboards with the same eight shapes on the turkey's feathers.

Materials

- Tom Turkey Gameboard (page 139) and Turkey Feather Patterns (page 140)
- light brown paper and an assortment of light-colored paper

Preparation: Make 4 copies of the Tom Turkey Gameboard on light brown paper. Make 4 copies of the Turkey Feathers on assorted light colors of paper. Laminate the papers and then cut apart the feathers. These will be the playing pieces.

Playing the Game

1. Each player selects one gameboard and places it on the table in front of him or her.
2. Turn all of the feathers facedown on the game table.
3. Select who will go first to turn over a feather. He or she will say the name of the shape that is on the feather and place it on top of the corresponding shape on his or her turkey.
4. The next player will also select a feather, say the name of the shape, and match the shape on the feather with a shape on his or her turkey.
5. Play continues until a player gets a feather with a shape that he or she has already collected. He or she identifies the shape and then replaces the shape facedown on the table so the next player can have a turn.
6. The next player tries to remember where the shapes are so that he or she knows where to fill in all the feathers on the turkey. (This is similar to the game Concentration.)
7. The player who fills in all the feathers on his or her turkey is the winner, or play can continue until all have filled in their turkeys.

Triangle Game

4 Players

Developing Skill: Players will need to identify the number on the die and place the corresponding number of erasers on the triangle spaces on the gameboard.

Materials

- 1 standard die (dots 1–6)
- 1 triangle-shaped delivery box (from a pizza or sub shop)
- contact paper and construction paper
- novelty erasers or buttons (enough to fill the grid)

Preparation: Cover the box with contact paper. Use the box to hold the game pieces. Cut out 4 different colored construction-paper triangles the same size as the box. Divide the triangle papers into triangle-shaped grids with each grid space large enough to place an eraser. Collect enough novelty erasers to fill the spaces on all four grids.

Playing the Game

1. Each player selects a gameboard and places it in front of him or her.
2. The players take turns rolling the die to see how many erasers they may select to cover the triangle spaces on their gameboard. When demonstrating this game, use phrases such as, "I'm covering three triangles" to encourage the players to use the new vocabulary.
3. Play continues until everyone has covered all the spaces on the triangles. Use the term "triangle" as often as possible during play to reinforce the vocabulary.

Use with Classroom Activities #1 and #6 on page 132.

Flat Shape Patterns

Rectangle

Semicircle

Use with Classroom Activities #1 and #6 on page 132.

Flat Shape Patterns *(cont.)*

Square

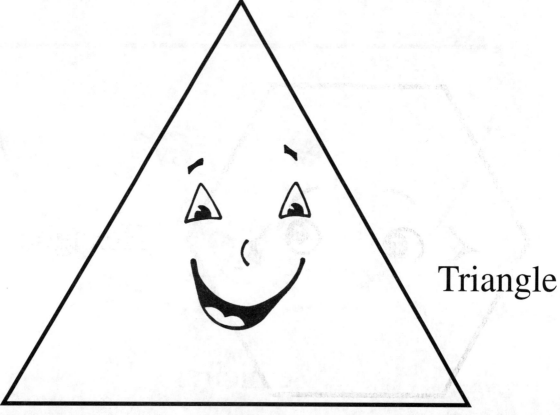

Triangle

136   ©Teacher Created Resources, Inc.

Use with Classroom Activities #1 and #6 on page 132.

Flat Shape Patterns *(cont.)*

Oval

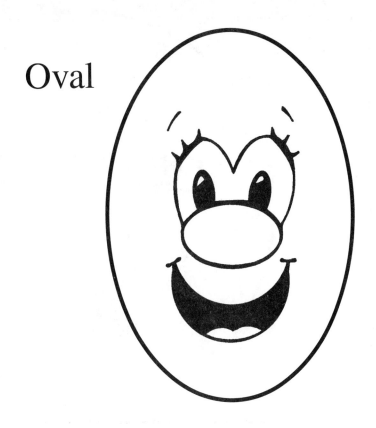

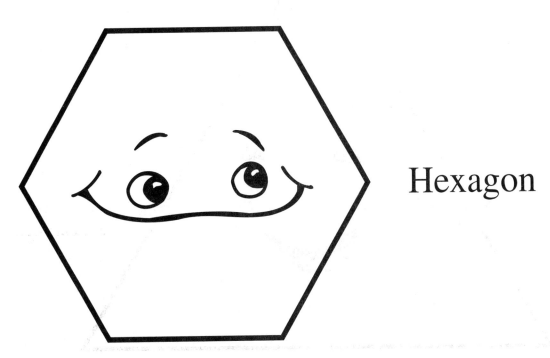

Hexagon

Flat Shape Patterns *(cont.)*

Circle

Rhombus

Use with the Tom Turkey Game on page 134.

Tom Turkey Gameboard

Turkey Feathers

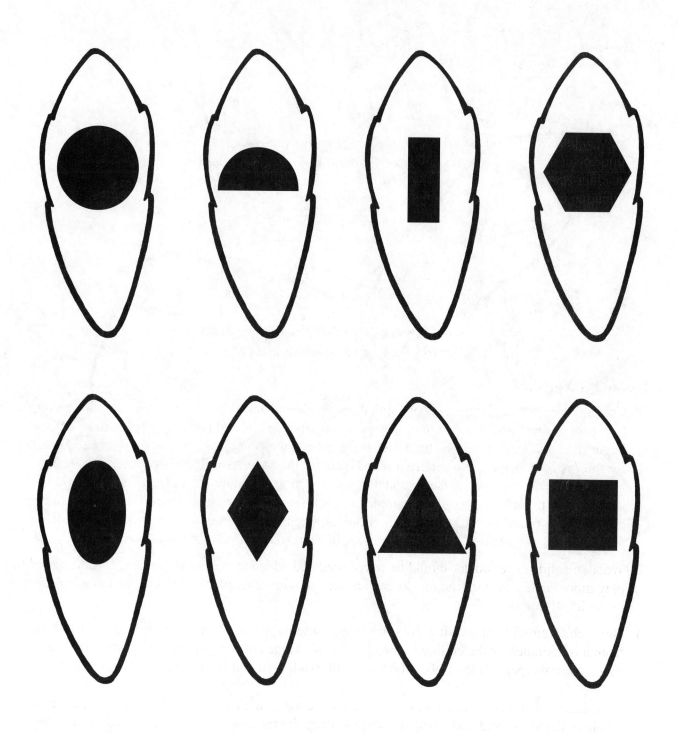

140

Solid Shapes

Educational Objectives: The child will be able to identify, name, and describe six solid shapes. He or she will know how to use vocabulary to describe the following three-dimensional terms: *sphere, cube, cylinder, cone, prism,* and *pyramid.*

Vocabulary:

 Sphere—a solid shape that looks like a ball

 Cube—a solid shape with six equal square sides (It is acceptable to call it a "box.")

 Cylinder—a solid shape that has a circle shape on each end and the sides go round and round

 Cone—a solid shape with a circle shape on one end that gradually changes to a point on the other end (it is acceptable to call it an "ice-cream cone")

 Prism—a triangular prism is a solid shape that has a triangle on each end and three rectangles on the sides

 Pyramid—a solid shape that has a square on the bottom and four triangles on the side that meet at a point

Classroom Activities

1. Use tagboard to make three sets of solid shape models. Make a small, medium, and large sample (see pages 146–149). Reduce the pattern for a smaller model and enlarge the pattern to get a larger model. Use a film canister, toilet-paper tube, paper-towel tube and/or an oatmeal container for the cylinder shapes, and different-sized balls for the sphere shapes. Cover each model in attractive paper. One set of these models can be used to introduce the new concept of solid shapes. Try to include everyday examples of each shape as it is introduced. For instance, a cube shape can be found in "cheese cubes." Store the wrapped shapes in a large shopping bag or surprise box. (Craft stores also sell a variety of Styrofoam shapes.)

2. Wooden solid shape blocks should be introduced and added to the Block Center. Children tend to play more creatively with the blocks in conjunction with other blocks, rather than isolating them in the Math Center.

3. Older children will enjoy using the solid shape patterns to make ornaments for a tree or to hang from a tree branch or the ceiling in the classroom. Make colored copies of the different solid shape patterns (pages 146–149). Show the children how to cut out the patterns and make the ornaments. Demonstrate how to fold the papers on the dotted lines and put glue on the tabs. Hold each glued piece and count to 10 or 20 while the glue takes hold or use a glue stick. Punch a hole in the shape and add a bright ribbon to hang the ornament. Use a paintbrush to add dots of glue to the sides of the solid shapes. Place the ornament into a plastic jar with about two tablespoons of glitter. Screw the lid tight, shake the jar, and count to a specific number. Take out the ornament and let it dry. If the folding is a difficult skill for the children, an aide or parent helper can help with this project. It is fun to have each child fill in a graph indicating which shape he or she chose to make. Review the results.

Solid Shapes *(cont.)*

Classroom Activities *(cont.)*

4. Make a kazoo from a toilet-paper tube. Use a 4" x 6" (10 cm x 15 cm) piece of construction paper. Cover the paper with glue and attach it around the tube. Decorate the kazoo with stickers. Cut a 4" (10 cm) square from a piece of wax paper. Place it over one end of the tube and secure it tightly with a rubber band. Show the children how to hum in the kazoo to make music.

5. Use a cone shape to make a craft. (*Example:* Use two cone shapes to make the body and head of an animal. Place one cone so it sits on the table. Turn the other cone upside down and staple it to the top of the first cone, creating a head. Add facial features and you have a cone-shaped animal [see diagram].)

6. Use the pyramid shape to wrap any small, light gift that you may be making for a family member. Transfer a large version of the pyramid shape onto a large piece of construction paper. (Cut off the gluing tab parts of the pattern.) Place the paper in a large box lid and marble-paint the paper using two or three different colors. (To marble-paint, use three different marbles and three different colors of paint. Place the marbles in a small amount of the paint. Remove each marble with a spoon and carefully place it in the box. Have the child gently roll the box so the marbles roll all over the paper creating a beautiful design. You can also use golf balls for this project to get larger lines.) Let the paper dry and fold on the fold lines. Wrap the gift in a piece of tissue paper and place it in the center of the pyramid on the square part of the pattern. Punch a hole in the top of the triangle parts of the pyramid and lace together with ribbon. Add a tag and you have a "Pyramid Present."

7. Make a "sphere" treat to enjoy. (Check to make sure there are no food allergies before planning this activity.) The children will enjoy rolling the dough into sphere shapes to make tasty cookies called "Snickerdoodles." Ask the children to predict what will happen to the spheres when they cook them. See recipe below.

Snickerdoodles

1 cup softened shortening

1½ cups sugar

2 eggs

2¾ cups flour

2 teaspoons cream of tartar

1 teaspoon baking soda

¼ teaspoon salt

1½ teaspoons cinnamon

2 tablespoons sugar

Mix softened shortening, 1½ cups sugar, and eggs. Stir in flour, cream of tartar, baking soda, and salt. Roll into balls the size of walnuts. Roll into mixture of 2 tablespoons sugar and 1½ teaspoons of cinnamon. Place 2 inches apart on an ungreased cookie sheet. Bake until lightly brown but still soft at 400°F for 8–10 minutes. Makes 5 dozen 2-inch cookies.

Solid Shapes *(cont.)*

Classroom Activities *(cont.)*

8. Decorate a miniature clown hat to take home and eat. Use ice-cream cones, frosting mix, and small candies. Invert the cones and cover them with the frosting. Let each child decorate as he or she wants. Send home in small paper bowls.

9. Introduce the children to the Giant Die Activity (see Classroom Activity #6 on page 155). The giant die is an excellent example of a cube shape.

10. Make a poster of "Solid Shapes in our World." Send a letter (page 145) home asking parents to help their children find pictures of solid-shaped items in magazines, catalogs, newspapers, etc. Have the children bring the pictures to school to add to a giant poster that can be displayed in the hallway. Make this a month-long project and let the children add more pictures as they bring them in.

11. Collect a used tennis ball or racquetball for each child to keep. Roll out the sphere shapes to your children to take home and play with. Have each child thank you for the sphere as he or she receives it (i.e., "Thank you for my sphere."). This helps him or her use the new shape vocabulary.

12. Make 3-D snowmen from dry laundry soap flakes. Pour soap flakes into a large bowl. Slowly add water. Use a hand mixer to mix the flakes and water to a consistency that can be molded by hand into sphere shapes. Have children make a large sphere and a small sphere. Hold them together by pushing one-half of a craft stick into the large sphere. Carefully slide the second, smaller sphere on top. Add facial features with buttons, pipe cleaners, small aquarium rocks, googly eyes, etc. Cut small pieces of fabric to use for a scarf. Add buttons. A cute hat can be made with a milk-jug cap or by cutting a top hat out of black tagboard.

13. Use the book *Brown Bear, Brown Bear, What Do You See?* by Bill Martin, Jr. for the model of a Class Shape Book. Take a picture of each child in the class holding a flat or solid shape. If you have a large class, use different colors for the shapes so every child will have a page in the book. Each page will have a child's name and the shape and color being held. (*Example:* "Teacher, Teacher, what do you see? I see Abbey holding a red prism looking at me." The next page will say, "Abbey, Abbey, what do you see?" This page is followed by "I see Beth with a blue pyramid looking at me.") The children will love looking at the book while they are reviewing shapes, colors, and classmates' names.

14. Create a "Cubes for Caring" bulletin board. Purchase small cube-shaped boxes from a box store. Have one box for each child in the class. Label each box with a child's name and/or picture and attach the boxes to the bulletin board. Explain that the cube-shaped boxes are for the children to give little surprises to their classmates. Send home a parent letter (page 145) to explain this activity. It is important that the teacher also participates in this fun activity. The boxes can be used over and over each year.

Solid Shapes *(cont.)*

Listening Activities

1. Place the wrapped solid shapes in a large shopping bag or surprise box. Let each child select and identify a solid shape. Then, ask the child to give the selected shape to a specific classmate. Have him or her describe the action. For example, "I'm giving a prism to Mindy."

   Work on the use of prepositions and review the centers in the classroom by asking the child to place the shape he or she has identified in a specific place. (*Example:* "Can you put your prism *next* to the Block Center?") Try to use *up, down, over, under, top, bottom, inside, outside, in front of, behind, between,* and *next to.*

2. To help make shape recognition more a part of the child's life, use the Solid Shapes Shopping Cards (pages 150–153). Copy the cards, cut them out, glue them onto construction paper, and laminate for durability. Display 5–6 different cards on an easel so they can be seen easily. For the listening game say, "We're going shopping for my uncle and I found the perfect gift. It is in the shape of a cube. Can you find the gift I bought for my uncle?" Make up a different sentence for each child describing a different gift and add a new card so each child will have 5–6 cards to choose from. Ask the child to identify the gift when he or she selects a card.

Selected Literature

Brown Bear, Brown Bear, What Do You See? by Bill Martin, Jr. (Philomel, 1999)

Captain Invincible and the Space Shapes by Stuart Murphy (HarperTrophy, 2001)

Cubes, Cones, Cylinders, & Spheres by Tana Hoban (Greenwillow Books, 2000)

Shapes by Karen Bryant Mole (Gareth Stevens Publishing, 2000)

Hello,

This month we will be introducing these six solid shapes to our students:

- Cone
- Cylinder
- Pyramid
- Cube
- Prism
- Sphere

We will be working on a class project to make a large poster titled, "Solid Shapes in our World." Please assist your child in finding pictures of the six solid-shaped items we are studying in the newspaper, catalogs, and ads. Help him or her cut them out and bring the pictures to school so the pictures can be added to our giant poster.

Some examples would be a telescope (cylinder), basketball (sphere), building block (cube), cookie jar (cylinder), clown's hat (cone), etc. Encourage your child to identify everyday items by their solid shapes.

Thank you for helping us make this an exciting learning experience.

Sincerely,

_____

Hello,

This month we will be learning about solid shapes. To help the children learn to use the new vocabulary, we are starting a fun activity called "Cubes for Caring." Our bulletin board will be covered with cube-shaped boxes. Each cube will be labeled with a child's name. Please encourage your child to create messages that show he or she cares about all his or her friends in our class. For example, your child might draw a special picture, make a treat, or write a special note. Please make sure your child puts his or her name on whatever is put in a box. Your child may choose to make a few messages each night. When they are all complete, he or she can bring them to school and place them in the cube-shaped boxes. We are hoping this activity will promote the idea of taking that extra step to show how we care for one another. Thank you for your assistance.

Sincerely,

_____

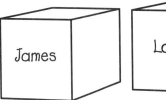 

James   Lori

Solid Shapes

Use with Classroom Activities #1 and #3 on page 141 and Listening Activity #1 on page 144.

Cube Pattern

Assembly Directions

1. Cut along the solid lines.

2. Fold along the dashed lines.

3. Tape or glue the cube together.

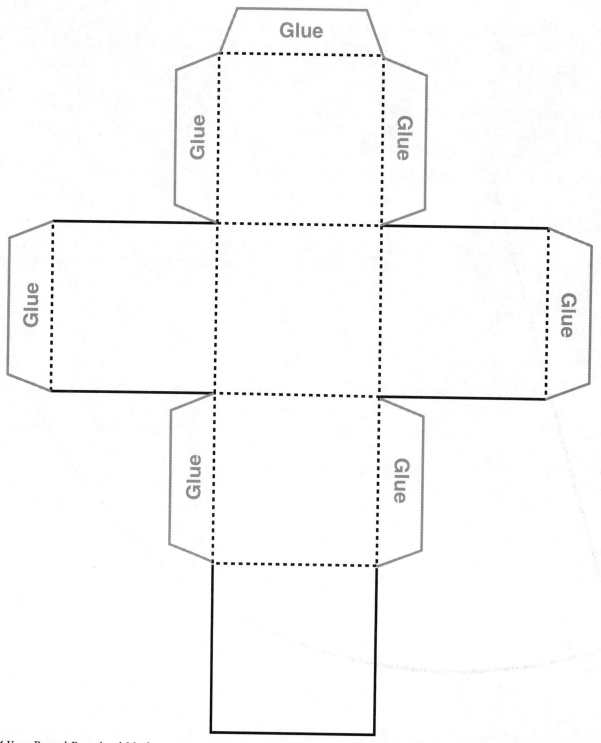

Use with Classroom Activities #1 and #3 on page 141, and Listening Activity #1 on page 144.

Cone Pattern

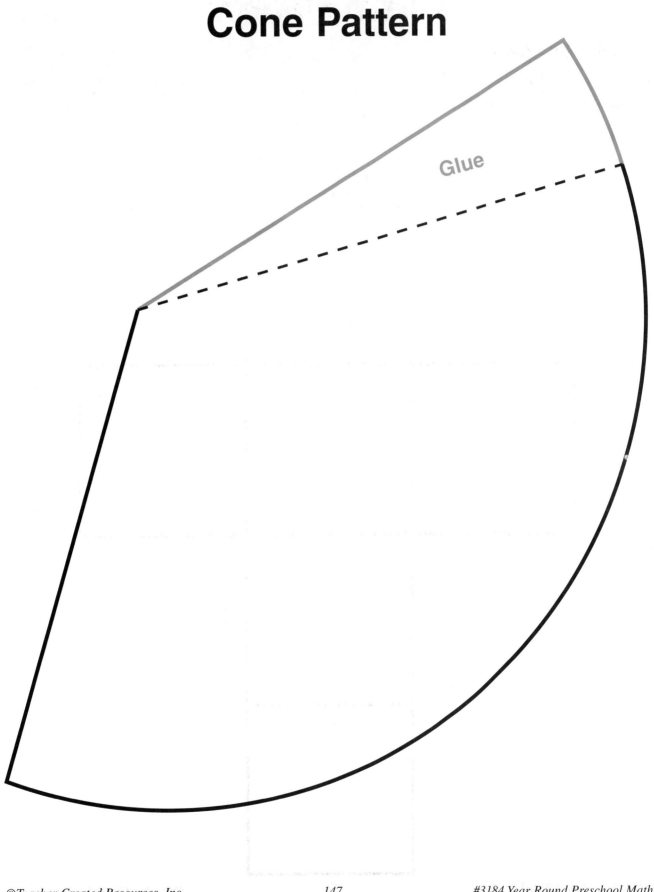

Glue

Use with Classroom Activities #1 and #3 on page 141 and Listening Activity #1 on page 144.

Prism Pattern

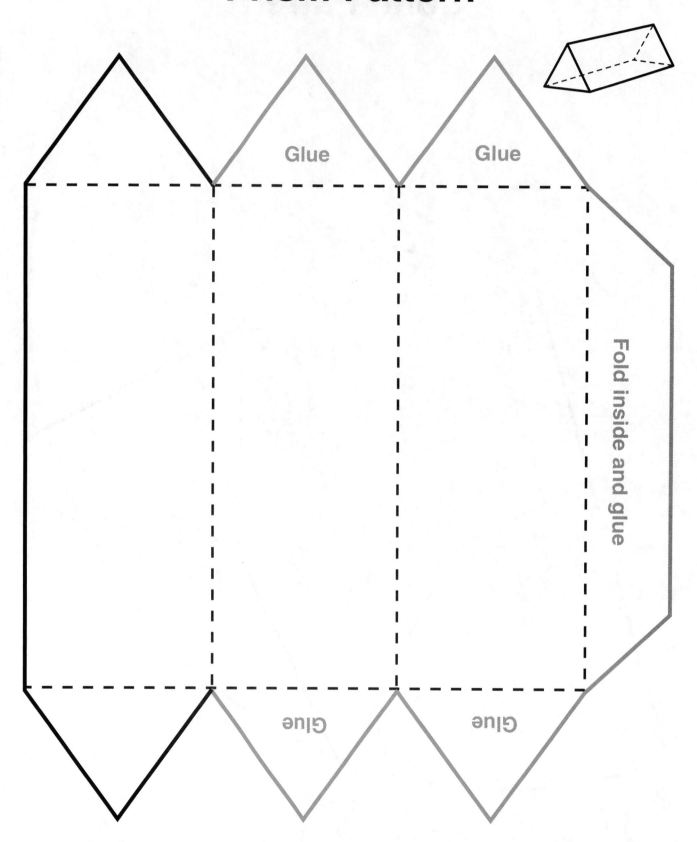

Use with Classroom Activities #1 and #3 on page 141 and Listening Activity #1 on page 144.

Pyramid Pattern

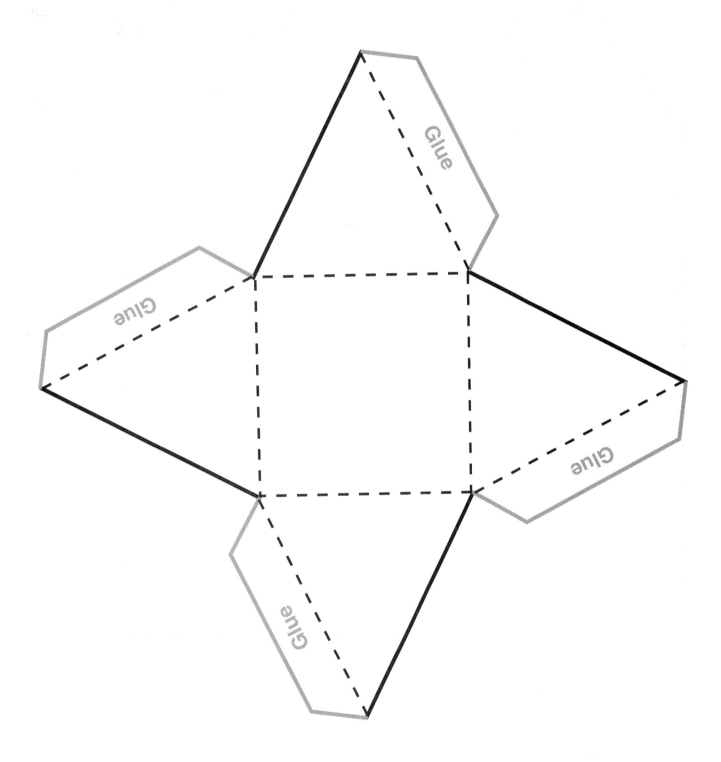

Solid Shapes Shopping Cards

Use with Listening Activity #2 on page 144.

Use with Listening Activity #2 on page 144.

Solid Shapes Shopping Cards *(cont.)*

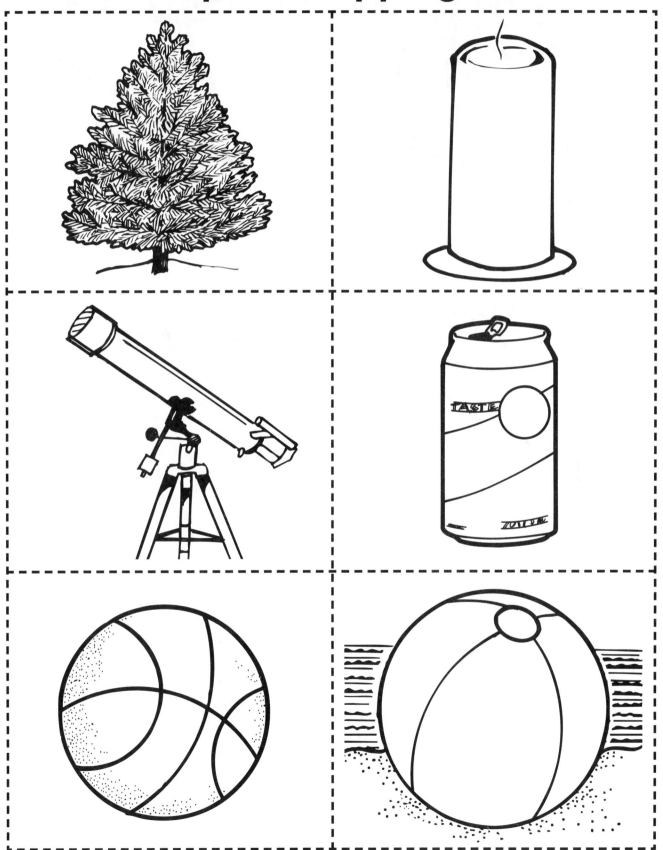

Use with Listening Activity #2 on page 144.

Solid Shapes Shopping Cards *(cont.)*

152

Use with Listening Activity #2 on page 144.

Solid Shapes Shopping Cards *(cont.)*

Cardinal Numbers 0–10

Educational Objectives: The child will be able to model, identify, name, and write the cardinal numbers 0–10. He or she will be able to compare and order the numbers up to 10.

Vocabulary: *Model*—a copy or imitation of an existing object

Identify—to recognize as being the thing known, described, or claimed

Name—a word or phrase by which a thing or class of things is known

Write—to form or inscribe symbols on a surface

Compare—to examine in order to observe or discover similarities or differences

Order—the sequence or arrangement of things or events

Classroom Activities

1. Introduce the cardinal numbers and the correct way to write each numeral at circle time. Use the Number Poem (pages 160–161) to make a poster. Recite a line while demonstrating how to make each numeral using a chalkboard, dry-erase board, or large easel. It may be beneficial to teach one or two numerals at a time. Stress that when writing each numeral, the writer starts at the top and moves down.

   One way to reinforce making numerals correctly is to do "pudding writing." First have children wash their hands. Give each child a small paper plate with a star at the top of the plate. Spread a small amount of pudding on each plate. Instruct the children to place the plate with the star farthest away from them. Help the children find and use their pointer fingers. Each numeral will be started just under the star. Demonstrate the correct way to make a numeral and then let them practice making the numeral in the pudding. Let them spread out the pudding and try again or introduce another numeral. Let the children lick their fingers after each practice. At the end of this fun practice, let them clean up the entire plate with their fingers.

2. Use colorful tape to put a number line on the floor in the classroom. Space the numbers no more than 12" (30 cm) apart. Use this line to have the children jump the numbers, one through ten. This line can be used for many different activities described in this book.

3. Make a "Flip Box" to use for number recognition (see page 162). Put a selection of 2" x 3" (5 cm x 8 cm) tagboard cards with the box. On one side of the card, write a numeral 0–10. On the other side put the correlating number of dots. The child tries to identify the number. He or she can then check the answer by putting the card in the Flip Box. As the card travels through the box, it will flip to reveal the side of the card with dots and the child can check the answer.

4. Demonstrate the correct way to complete a dot-to-dot activity. It is best to use a dry-erase board or chalkboard so all the children can see well. Draw dot shapes (a simple, numbered shape or picture like a star or house). Emphasize that it is important to find the next number before the pencil or crayon begins to move. Remind the children to connect the dots and not the numbers. Add dot-to-dot activities to the Math Center. These can be colorful laminated samples that the children can fill in with oil crayons and then erase with a tissue, or actual paper and pencil examples that the children can take home. Review the pages with the children before they take them home. (See pages 163–166.)

Cardinal Numbers 0–10 *(cont.)*

Classroom Activities *(cont.)*

5. Enlarge the Numeral Cards (pages 167–168). Copy the cards onto tagboard, cut out, and laminate to create stencils. Put stickers on each stencil to correlate with the number represented. For example, put three ladybug stickers on the three stencil. (This also helps the child identify the correct way to place the stencil before he or she starts tracing it.) Make sure to let the children know this when the activity is introduced. Place stencils in the Art Center with colored pencils and paper so the children can trace them. Demonstrate this activity before setting it out.

6. Make a die for your class to use with the Giant Die activity (see page 170). Use a cube-shaped tissue box to make the die. Cut the box down so it will be an exact cube. Stuff the box with newspaper and reassemble. Cover the hole with a scrap of cardboard. Tape all the loose ends together with masking tape. Cover the box with solid-colored contact paper or bulletin board paper. Buy a supply of color-coded dots or stickers from an office supply store. Use them for the dots on the die. Use a commercial die to demonstrate that each dot is in a very special place on the die. Then show the children the completed homemade die. Make a copy of the Giant Die Game Sheet (page 169) and laminate it. Show the children the Giant Die Game Sheet. Ask for volunteers to demonstrate the six different skills that correlate to the numbers 1–6 on the die. For example, if the child rolls a 5, he or she must count out and do five Frog Jumps. It is important to count carefully and not do any more or less jumps than indicated on the die.

7. Provide assorted tactile items for the children to use to learn the numbers. Model the activities.

   • Make play dough tracers (stencils) using the Numeral Cards (pages 167–168). Copy and enlarge each numeral and glue it onto construction paper. Laminate the cards. Encourage the children to roll out snake shapes and cover the numerals with the dough. The children can also make a coordinating number of cookies, etc., to place next to each numeral card.

   • Put approximately one cup of bright paint into large plastic, resealable bags. Close the bags securely and reinforce with masking tape. Place the bag on the table and spread out the paint by pressing lightly on it. Place copies of the Numeral Cards (pages 167–168) for the children to copy next to the bags. Add a star at the top of each numeral. The children practice making the numerals by pressing on top of the bags using their pointer fingers. Remind the children that all numerals are started at the top. An alternative activity is to make several small bags with a numeral 0–10 written on the top of the bag with a permanent marker. The children trace over each numeral to practice the correct way to make the numeral.

   • Buy sheets of sandpaper in different grits. Cut each sheet in half and write a numeral (0–10) using a permanent marker. Add a star at the top of each numeral to remind the children to start at the top when writing numerals. Have the children trace over each numeral with their pointer finger or a cinnamon stick.

   • Set out food trays with thin layers of shaving cream, salt, corn meal, sand, cornstarch, or any other safe media with a good texture. (Use one texture at a time.) Show the children how to spread out the media and model how to practice making numerals the correct way. Have samples of the numerals available for the children to copy.

   • Let the children use a damp sponge to practice writing numerals on the chalk board.

   • Provide a large bin filled with dried beans, Styrofoam pieces, rice, etc. Hide magnetic numerals in the bin for the children to find. Have the children identify the numerals as they find them.

Cardinal Numbers 0–10 *(cont.)*

Classroom Activities *(cont.)*

8. Make large construction paper numerals to decorate your classroom. Several children will work together on each numeral. Have the children choose a different medium to cover each numeral. Children tend to take ownership of the numeral that they create, so it is good to match up children with a numeral they have not yet mastered. Cover the numerals with pasta, pompoms, bottle caps, ribbons, curled construction paper, etc. Hang the finished numerals on the wall for all to enjoy.

9. Set up a Number Detective Station. Provide magnifying glasses, old hats, vests, clipboards, detective badges, pencils, and tally sheets in the Math Center (see pages 171–172). Before the children arrive at school, make copies of the Numeral Cards (page 167–168). Glue them onto construction paper squares and laminate. Hide the numerals around the room on the sides of shelves, chair legs, under tables, etc. Hide them so the children do not have to move anything to find them. The children will put the tally sheets on clipboards and use the magnifying glasses to find numerals throughout the room. Demonstrate how to do this activity and how to make a tally mark each time a numeral is found. Change the locations of the numerals and include a new numeral to learn each day this activity is offered. (Limit this activity to a few children each day. Then each child can keep his or her Number Detective Badge as it is earned.) Award each child with a special certificate (page 170) after completion of the activity.

10. Make a Magnetic Fishing Activity. Copy the Fish Patterns (page 173) several times and cut them out. Glue the fish onto construction paper. Add the matching number of dots on the back of each fish and laminate for durability. Staple several staples close together on the mouth of the fish to make a magnetic field. (Paper clips can also be used.) Use thin dowels to make fishing poles. Add a string with a doughnut magnet tied to the end. Cut a pretend pond from light blue bulletin board paper or tagboard. Place the fish on the pond and let the children enjoy fishing. Have them say the number as they catch each fish. If they are not sure, they can turn the fish over and count the dots. Have numbered baskets for children to place their fish in once they are caught.

11. Make an Ice-Cream Cone Matching Activity. Copy 10 Ice-Cream Cone Patterns (pages 176–178) on tan or brown construction paper. Copy each Ice-Cream Scoop Pattern (pages 174-176) on a different color of paper and laminate. Attach small pieces of Velcro® onto the front top of each cone. Attach the matching pieces of Velcro on the bottom back of each scoop of ice cream. Have the children match the appropriate cone with the corresponding number of chips in the ice cream. As the children find the matching cones, have them stick the matches together. After the cones have been put together, they can be placed in the proper order from left to right.

12. Before the children arrive at school, make several Invisible Numerals. Use a white crayon on white paper to color in large numerals. Place the numbers in the Art Center. Demonstrate how the numerals will appear if they are covered with watercolor paints. Encourage each child to identify the numeral he or she discovers.

Cardinal Numbers 0–10 *(cont.)*

Listening Activities

1. Children sometimes have trouble learning the concept of "0." Use this activity to help the children discriminate between 0 and 1. Ask a child to answer each question with a sentence using the number 0 or 1.

- How many noses do you have?
- How many tails do you have?
- How many times have you gone to the moon?
- How many heads do you have?
- How many belly buttons do you have?
- How many zebras live with you?
- How many tongues do you have?
- How many spots do you have?
- How many frogs are you holding?
- How many tummies do you have?
- How many mouths do you have?
- How many crayons are in your hand?
- How many clown shoes are you wearing?
- How many flowers are in your hand?
- How many watches am I wearing?
- How many hats are you wearing?
- How many heads do I have?
- How many pizzas are in your hand?
- How many giraffes are in our room?
- How many turtles are at our circle?

one!

2. Use a dry-erase board or chalkboard. Write a few of the numerals 0–10, leaving several blank spots. Ask each child what number is missing. Give a different problem to each child.

3. Ask number-related questions. For example, "What number comes between 5 and 7?" "What number is one more than 6?" "What number comes after 8?" etc. Provide a number line for the children to use if they need assistance. Ask each child to point to the number line. Encourage him or her to use a complete sentence when answering.

Cardinal Numbers 0–10 *(cont.)*

Listening Activities *(cont.)*

4. Play the Classroom Count Listening Game. Add your own questions to the suggestions below.

How many tables are in our classroom?	How many clocks are in our classroom?
How many windows are in our classroom?	How many 2's are on the calendar?
How many flags are in our classroom?	How many doors are in our classroom?
How many students are in our room today?	How many teachers are in our room now?
How many girls are here today?	How many boys are here today?
How many sinks are in our classroom?	How many (student's name) are here?
How many lights are in our classroom?	How many children are wearing _____?

Selected Literature

Anansi the Spider by Gerald McDermott (Henry Holt & Company, Inc., 1972)

The Baseball Counting Book by Barbara McGrath (Charlesbridge Publishing, 1999)

The Cheerios Counting Book by Barbara McGrath (Scholastic, Inc., 1998)

City by Numbers by Stephen John (Puffin Books, 2003)

Eight Silly Monkeys by Steve Haskamp (Intervisual Books, Inc., 2003)

Feast for 10 by Cathryn Falwell (Houghton Mifflin, 1995)

Five Little Monkeys Sitting in a Tree by Eileen Christelow (Houghton Mifflin, 1999)

Gray Rabbits 1, 2, 3 by Alan Baker (Houghton Mifflin, 1999)

I Spy Two Eyes by Lucy Micklethwait (William Morrow & Co., 1998)

More Than One by Miriam Schlein (Greenwillow Books, 1996)

My Truck is Stuck by Kevin Lewis (Hyperion Books for Children, 2002)

The Numbers by Monique Felix (Creative Teaching Press, 1993)

Pigs from 1 to 10 by Arthur Geisert (Houghton Mifflin, 1992)

Seven Blind Mice by Ed Young (Puffin Books, 2002)

The Seven Silly Eaters by Mary Hoberman (Harcourt, 1997)

Six-Dinner Sid by Inga Moore (Alladin Books, 1993)

Ten Minutes Till Bedtime by Peggy Rathmann (Puffin Books, 2004)

When Sheep Cannot Sleep by Satoshi Kitamura (Farrar, Straus & Giroux, 1988)

Valentine Tweezers   **2–4 Players**

Developing Skill: Players will need to be able to recognize the number on the die and count out the correct number of pompoms. Players will also need to have the dexterity to manipulate a tweezer to pick up a pompom and understand one-to-one correspondence to place one pompom on each grid space.

Materials

- 4 plastic tweezers and assorted small pompoms
- 1 standard die (with numerals 1–6)
- 1 heart-shaped candy box
- construction paper

Preparation: Cut out 4 different heart-shaped pieces of construction paper the same size as the box. Divide the paper into grids approximately 1" (2.54 cm) in size. Laminate the papers for durability.

Playing the Game

1. Each player selects a gameboard and a set of tweezers.

2. Place all the pompoms inside the heart box.

3. Select which player will go first. He or she will roll the die and use the tweezers to pick up the appropriate number of pompoms and place them on the grid shapes.

4. Play continues until a player fills all the spaces on his or her grid.

Musical Numbers   **Circle Time Game**

Developing Skill: Players need to be able to identify the numerals 0–10 and know how to place them in the correct order.

Materials
- Numeral Cards (pages 167–168)

Playing the Game

1. Give each player one numeral card.

2. Turn on some music and let the players dance.

3. After a short period of time, stop the music.

4. When the music stops, the players have to line up in the correct order. Show the players where they need to start the number line to ensure that it will go from left to right.

5. Younger players will enjoy this activity by finding a copy of the numeral they are holding on the floor and standing on it. (These numerals can be the ones on the floor number line.)

Use with Classroom Activity #1 on page 154.

Number Poem

0	Start at the top around I go, that is how I make the number zero.
1	A straight line down, it is easy and it is fun, that is how I make the number one.
2	A curve and a line will be the clue, that is how I make the number two.
3	A curve for you a curve for me, that is how I make the number three.
4	Down and over, down some more, that is how I make the number four.

Use with Classroom Activity #1 on page 154.

Number Poem *(cont.)*

5	First a line, then take a dive, around I go to make the number five.
6	Down to the ground to pick up sticks, around I go to make the number six.
7	Across the sky and down from heaven, that is how I make the number seven.
8	Make a slanted line but do not wait, back I go to make the number eight.
9	Draw a circle and then a line, that is how I make the number nine.

Use with Classroom Activity #3 on page 154.

Flip Box Directions

Materials

- half-gallon milk or juice container
- masking tape/clear tape
- contact paper, wrapping paper, or construction paper
- tagboard
- ruler

Procedure

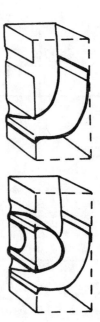

1. Pull open the top of the container. Make two measurements down the front side of the container. The first one should be at 1½" (3.8 cm) and the second one should be at 2½" (6.3 cm).

2. Use scissors to remove the 1" (2.54 cm) space in between these marks. Flip the container and mark the bottom of the box in the same manner. Once again cut out the 1" (2.54 cm) space in between the lines.

3. Mark and cut out a piece of tagboard measuring 6½" (16.3 cm) by 3½" (9.2 cm). Fold down ½" (1.3 cm) on both ends. Mark and cut out another piece of tagboard measuring 7½" x 3½" (19.3 cm x 9.2 cm). Fold down ½" (1.3 cm) on *only* one end of this strip. Carefully push the unfolded end of the longer strip into the bottom opening of the container. The folded flap will end up resting on the bottom part of the container. (See diagram.) Use tape to attach this securely. This piece of tagboard creates a slide inside the box.

4. Carefully push the smaller strip into the bottom opening and back out the top opening. This time the bottom flap will be attached to the top of the bottom opening and the top flap will be attached to the bottom of the top opening. (See diagram.) Secure with tape.

5. Because of slight changes in box sizes, try a test card at this point. Cut tagboard into several 2½" x 3" (3.3 cm x 8 cm) pieces to use as flip cards. Place one card in the top opening of the box and give it a slight push. The card should travel through the box and come out the bottom opening. If the card does not travel through the flip box, check to make sure the two pieces of tagboard are not touching. If they are, push the slide piece of tagboard either up or down to allow enough space for the test card to travel through the box smoothly. Once a successful run has been completed, tape the top end of the slide piece of tagboard securely to the back of the box.

6. Fold down the top flaps of the container that were initially pulled open. Try to make a smooth, flat top. Cover the entire outside of the box with the paper of your choice. Label the top of the box with words or a picture so the children will know the correct way to use the Flip Box.

Flip Cards can be used in a variety of ways. For instance, the box can be used for number recognition. A child looks at one side of the card (the numeral 7) and says what he or she thinks it says. The card is then put into the top opening of the box. It will automatically flip to the other side of the card as it travels along the slide to reveal seven dots. The child can then count the dots to make sure he or she has named the correct number. It helps to put arrows on the face of the card to indicate which way to insert the flip cards into the box. Put the question on one side and the answer on the other. (*Note:* Put one trial card into the box first to make sure all the answers will be written facing the correct way.) Flip boxes can be used for addition and subtraction problems as well.

Use with Classroom Activity #4 on page 154.

Dot-to-Dot (1–10)

Dot-to-Dot (1–10) *(cont.)*

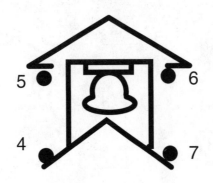

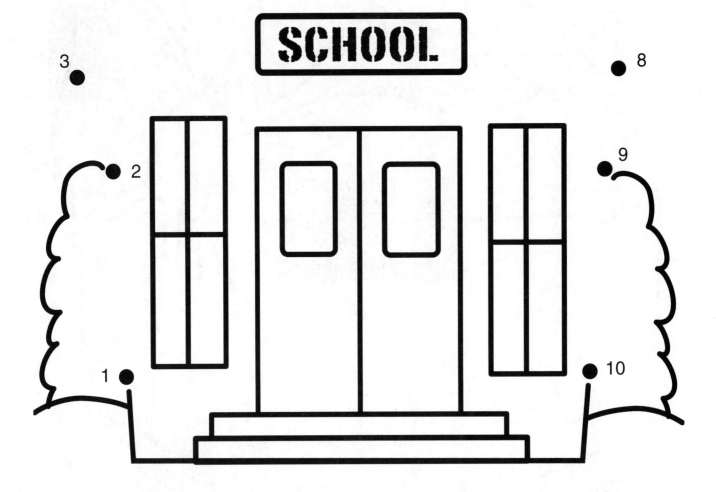

Use with Classroom Activity #4 on page 154.

Dot-to-Dot (1–10) *(cont.)*

Dot-to-Dot (1–10) *(cont.)*

Use with Classroom Activities #5, #7, and #9 on pages 155 and 156 and the Musicial Numbers game on page 159.

Numeral Cards

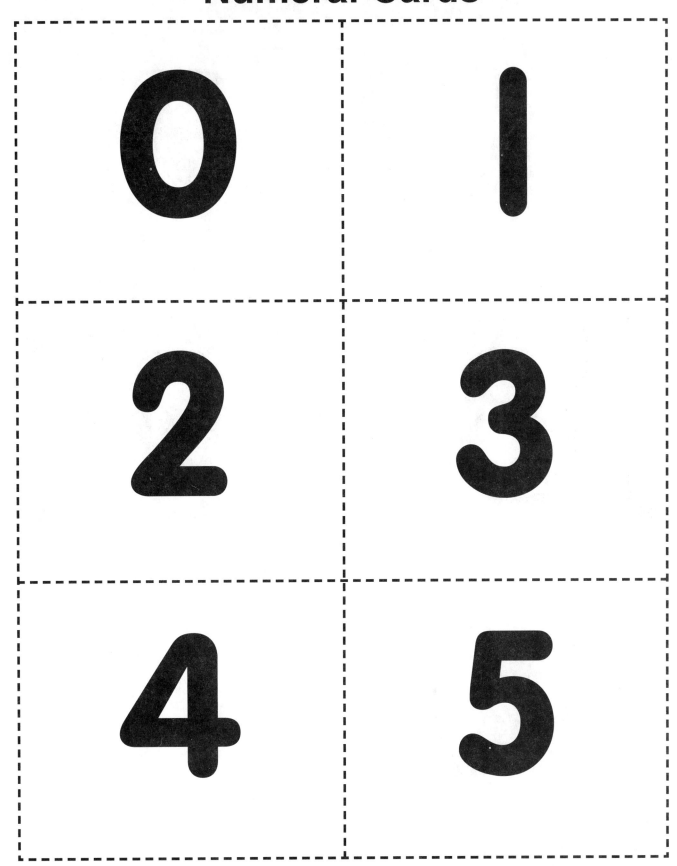

Use with Classroom Activities #5, #7, and #9 on pages 155 and 156 and the Musicial Numbers game on page 159.

Numeral Cards *(cont.)*

Use with Classroom Activity #6 on page 155.

Giant Die Game Sheet

	1 **Somersault**
	2 **Log rolls**
	3 **Jumping jacks**
	4 **Hops on one foot**
	5 **Frog jumps**
	6 **Jumps on two feet**

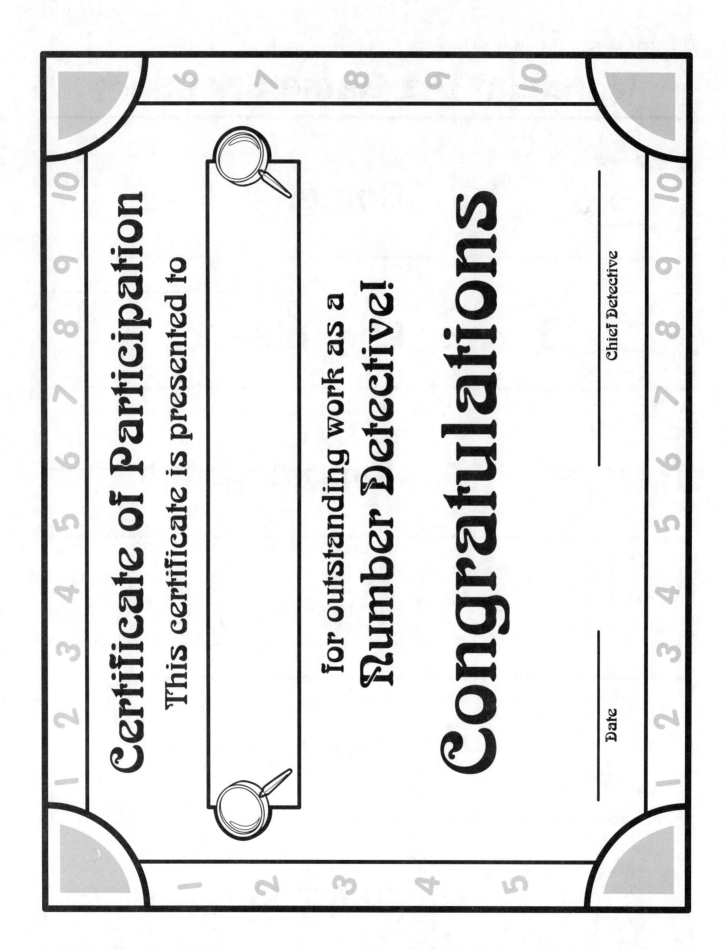

Certificate of Participation

This certificate is presented to

for outstanding work as a
Number Detective!

Congratulations

_____ Chief Detective

_____ Date

Cardinal Numbers 0–10

Number Detective Tally Sheet

0	1	2
3	4	5
6	7	8
9	10	Write a new numeral.

0	1	2	3	4	5
6	7	8	9	10	

Number Detective Badges

Use with Classroom Activity #10 on page 156.

Fish Patterns

Ice-Cream Scoop Patterns

Use with Classroom Activity #11 on page 156.

Ice-Cream Scoop Patterns *(cont.)*

Ice-Cream Scoop & Cone Patterns

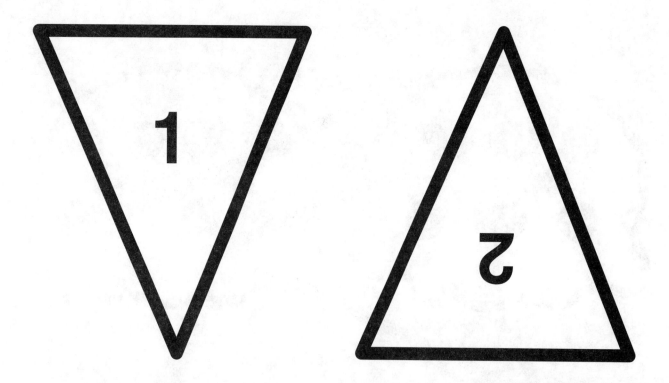

Use with Classroom Activity #11 on page 156.

Ice-Cream Cone Patterns

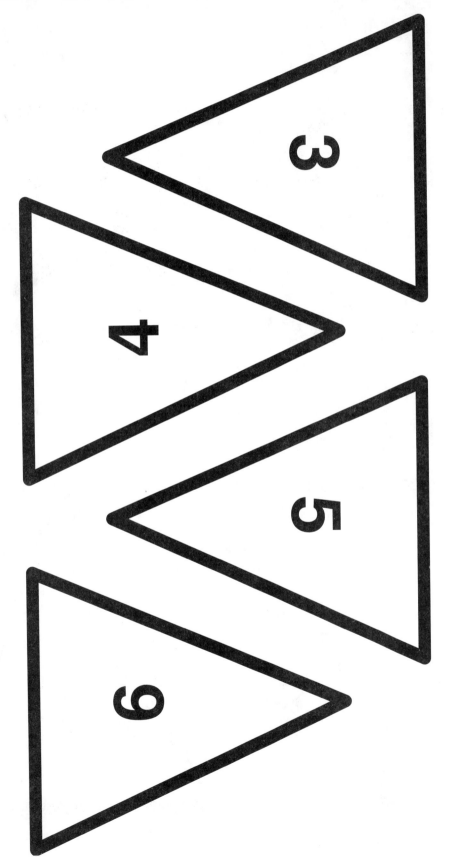

Ice-Cream Cone Patterns *(cont.)*

178

Pairs

Educational Objectives: The child will be able to identify and name items that are commonly presented as a pair. The child will be able to count by two's with the assistance of a number line.

Vocabulary: *Pair*—two similar or corresponding things joined, associated, or used together

Classroom Activities

1. Introduce the concept of pairs with a colorful pair of socks or gloves. Explain that items that come in pairs are used together. Explain that later in the day the class will have a Sock Hunt. Collect pairs of colorful socks. Hide the socks while the children are out of the room and have a Sock Hunt when they return. Then have them work together to get all the pairs matched together. *Variation*: Hide only one sock of each pair. Give one sock to each child and have him or her find the mate.

2. Send home a letter (page 181) asking the children to bring in a pair of something to school. Let each child introduce the items he or she has brought to school at circle time.

3. Introduce the concept of counting by twos. Some people call this "skip counting." Demonstrate how to jump by twos on a number line. Encourage the children to practice this during free choice time. Count by twos often throughout the year.

4. Model how to do a dot-to-dot activity that is completed by counting by twos (see page 182).

5. Make a poster of pairs using the pictures on pages 183–184. Place it in the Math Center.

6. Have children divide their snack into piles of twos, or pairs, and practice counting by twos.

7. Teach the class the old cheer: *"Two, four, six, eight, who do we appreciate?* (Name of child or teacher, name repeated,) *Yea!"* You can also teach: *"One, three, five* (pause), *seven, nine. Who do we think is mighty fine?* (Name of child or teacher, name repeated), *Yea!"*

8. Demonstrate how to use a calculator to count by twos. Use a calculator with a large viewing window. Turn it on and press + 2 = . Keep pressing = and the window will automatically display each number in intervals of two. Let each child have a turn using the calculator during the day.

9. Plan a unit on animal tracks. Look at the differences in the front and back pair of prints on some animals. Make copies of the Animal Footprint Patterns (pages 185–186). Let the children make a poster or a book of common animal tracks in their part of the country.

10. Ask your children if their footprints would be a pair. Explain that they are a unique pair because no one else will make a footprint just like the one they make. Have the children take off their shoes and socks and make a special pair of footprints to take home. It helps to use a paint roller to apply an even coat of paint. Supply a dishpan with warm, soapy water to clean the feet afterwards. (This activity can also be done with handprints.)

11. Add a set of "Counting Shoes" to the Math Center. Make copies of the Counting Shoe Patterns (pages 187–189). Add some color, glue the shoes onto tagboard or file folders, and laminate them. Punch a hole in the corner of each shoe and use a metal ring to attach the shoes in the correct order. Demonstrate how to use these fun shoes to practice counting by twos.

12. Make a pair of matching mittens. Provide a mitten pattern for the children to trace and cut out. (You can make a large copy of the Mitten Pattern on page 183.) Let the children have fun painting the two mittens so they match and string them together. Try marble-painting, print transfers, making stripes with a small roller, spatter-painting, etc.

Pairs *(cont.)*

Listening Activities

1. Collect enough pairs of items for each child to choose a different item. Put one half of the pairs in one large container such as a laundry basket. Put the other items in a different mystery box or bag. Let each child reach into the mystery box and pull out one of the items. Have the child explain what the item is that was chosen and have him or her find the matching item in the laundry basket.

2. Use the Matching Pairs Concentration Cards (pages 183–184). Place one card from each pair on an easel so the cards can be easily seen. Place the other cards on the carpet. Let each child select and identify one item from the easel and then find the matching item to make a pair.

3. Ask each child to demonstrate how to count by twos by jumping on a number line. Let each child select if he or she wants to start by jumping from number 1 or from number 0. (The teacher may need to hold the child's hands the first time.) Have each child count aloud as he or she jumps on each number.

Selected Literature

Alligator Shoes by Arthur Dorros (Puffin Books, 1992)

Body Pairs by Lola Schaefer (Heinemann Library, 2003)

The Counting Race by Margaret McNamara (Aladdin, 2003)

Counting Sheep by Julie Glass (Random House Books for Young Readers, 2000)

The Crayon Counting Book by Pam Munoz Ryan (Charlesbridge Publishing, 1996)

Eating Pairs by Sarah Schuette (Capstone Press, 2003)

Footprints in the Snow by Cynthia Benjamin (Scholastic, Inc., 1998)

Hello Toes! Hello Feet! by Ann Paul (DK Publishing, Inc., 1998)

How Many Birds? by Don Curry (Capstone Press, 2000)

How Many Feet in the Bed? by Diane Hamm (Aladdin, 1994)

The Missing Mitten Mystery by Steven Kellogg (Puffin Books, 2002)

New Shoes for Silvia by Johanna Hurwitz (William Morrow & Co., Inc., 1993)

One Mitten by Kristine George (Houghton Mifflin, 2004)

Red Lace, Yellow Lace by Mike Casey (Barron's Educational Series, 1996)

Shoes, Shoes, Shoes by Ann Morris (William Morrow & Co., 1998)

Spunky Monkeys on Parade by Stuart Murphy (HarperTrophy, 1999)

Two of Everything by Lily T. Hong (Albert Whitman & Co., 1993)

Two Ways to Count to Ten by Ruby Dee (Henry Holt & Company, 1991)

What Comes in 2's, 3's, and 4's? by Suzanne Aker (Aladdin, 1992)

Whose Shoe? by Margaret Miller (HarperCollins, 1991)

Matching Pairs Concentration　　　　　　**2–4 Players**

Developing Skill: Players will need to be able to identify matching pairs of objects. They will need to remember where similar items are so they can collect matching pairs.

Materials

- Matching Pairs Concentration Cards (pages 183–184)

Preparation: Make two copies of the cards. Glue them onto construction paper and laminate for durability.

Playing the Game

1. Line up all the playing cards in rows, facing down.

2. Select who will go first. This player turns over two cards and places them so everyone can see them. If they match, he or she gets to keep them and takes another turn. If not, the cards are turned over and the next player gets to go. As more cards are turned over, it gets easier to find a match. Play continues until all the pairs have been matched up.

3. The player with the most matching sets wins the game.

Hello,

This month our class is learning about pairs. We will be teaching the children this concept by providing some school activities we are sure you will be hearing about! You can assist us by helping your child find an item at home that is an example of something that comes in pairs. He or she can bring it to school on _____ to share with the class. The list below will give you some suggestions. Please make sure that your child will be able to carry whatever item is brought to school. (No breakable items please.)

Suggestions:

- ✳ Socks
- ✳ Mittens/gloves
- ✳ Swimming flippers
- ✳ Shoes
- ✳ Salt/pepper shakers
- ✳ Skates

We look forward to a fun-filled day. Thank you for your cooperation.

Sincerely,

_____

Counting by 2s Dot-to-Dot

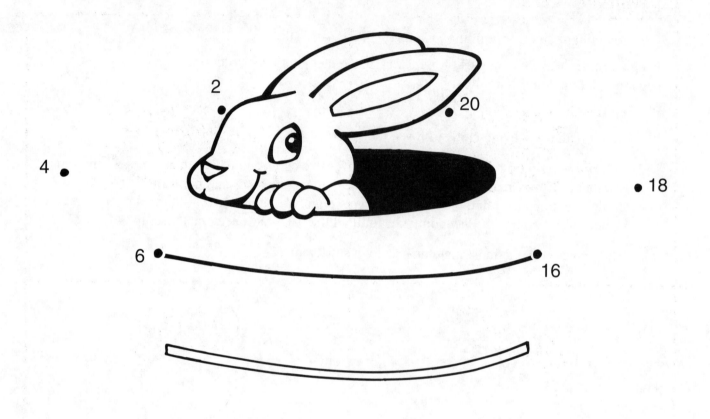

2

20

4

18

6

16

8

14

10

12

Use with the Matching Pairs Concentration game on page 181, Classroom Activity #5 on page 179, and Listening Activity #2 on page 180.

Matching Pairs

Concentration Cards

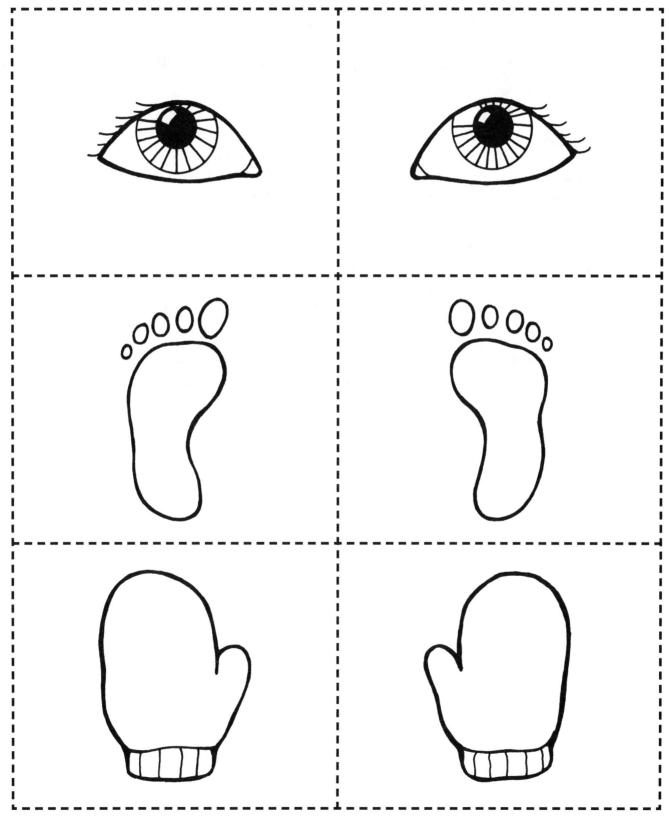

Pairs

Use with the Matching Pairs Concentration game on page 181, Classroom Activity #5 on page 179, and Listening Activity #2 on page 180.

Matching Pairs *(cont.)*

Concentration Cards *(cont.)*

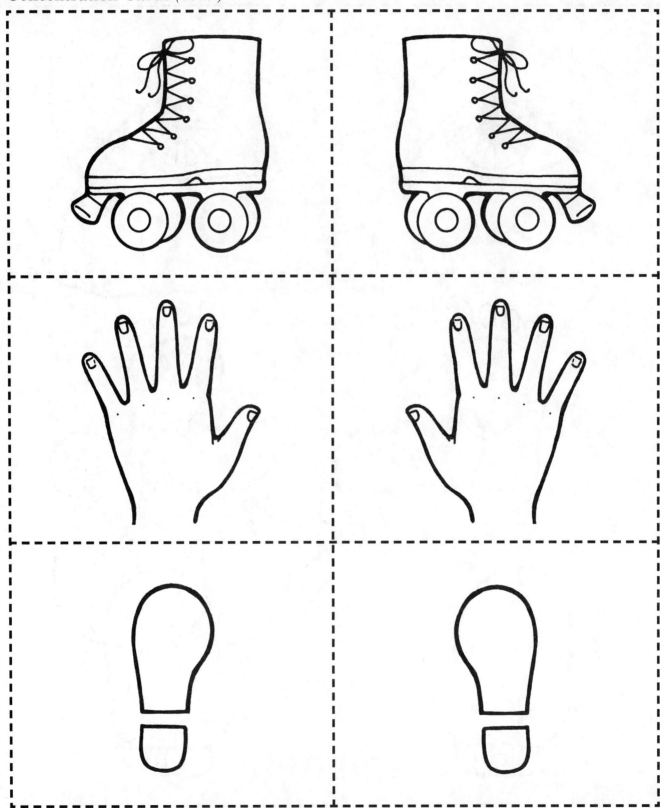

Use with Classroom Activity #9 on page 179.

Animal Footprint Patterns

rabbit

squirrel

Use with Classroom Activity #9 on page 179.

Animal Footprint Patterns *(cont.)*

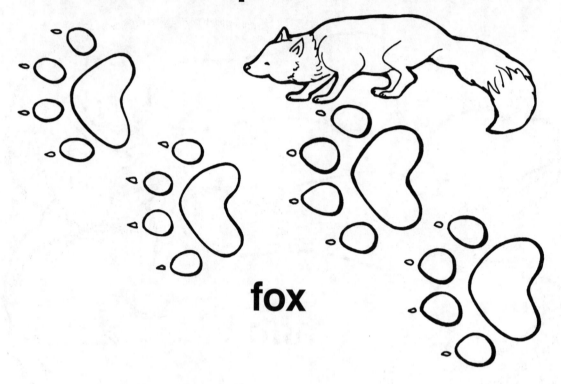

fox

deer

Use with Classroom Activity #11 on page 179.

Counting Shoe Patterns

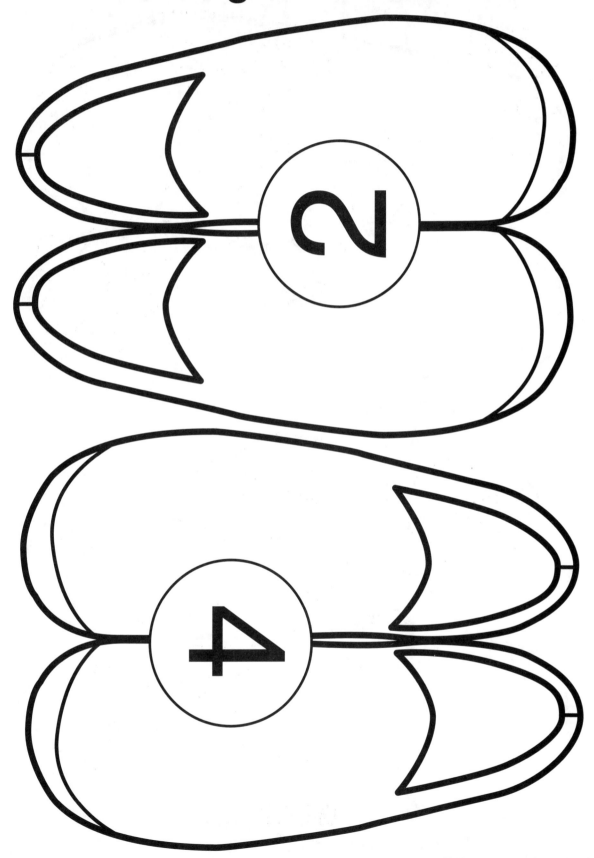

Counting Shoe Patterns *(cont.)*

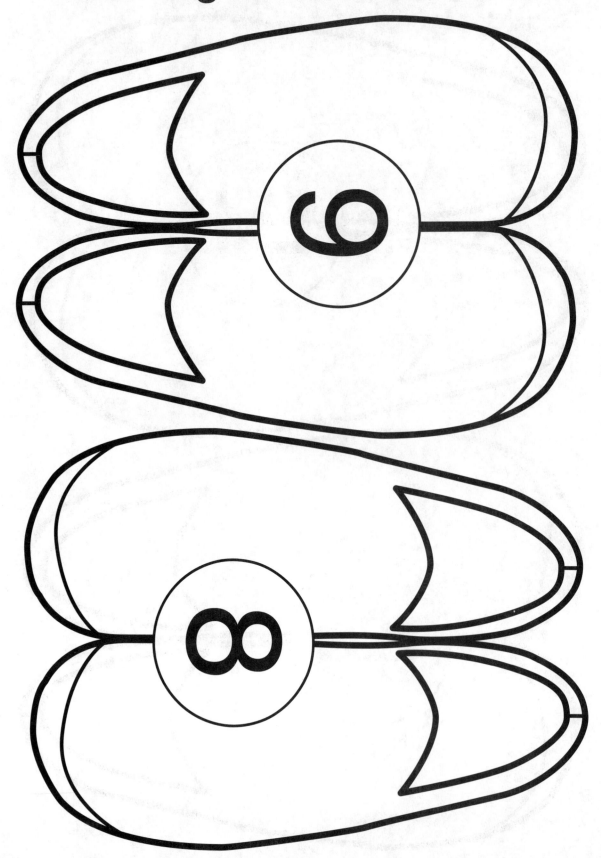

Use with Classroom Activity #11 on page 179.

Counting Shoe Patterns *(cont.)*

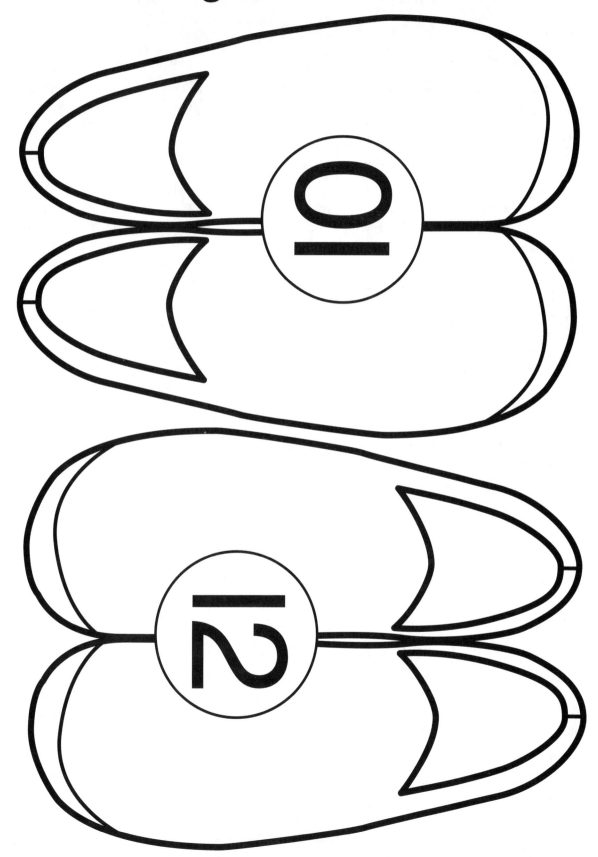

Counting Backward

Educational Objective: The child will be able to count backward from any given number between 10–0.

Vocabulary: *Backward*—doing something in a way contrary to the normal or usual way, in reverse

Classroom Activities

1. Plan a Backward Day. (This is fun to do around April Fool's Day.) Send home a letter (page 191) asking the parents to send the children to school wearing some of their clothes backward. Have the children sit at the table with the chairs turned backward, run through the class schedule backward, and do other activities backward that day. Read the book *Silly Sally* by Audrey Wood, which tells the story of a little girl enjoying a silly day.

2. Introduce the concept of counting backward at circle time by showing the children how to jump on the number line in reverse order. (Use the same number line on the floor that was introduced with Classroom Activity #2 on page 154.) Demonstrate how to jump from 10 down to 0. This can be done both facing forward and facing backward to make it more challenging. Encourage each child to try this new activity and say the numbers as he or she jumps. Dismiss each child from circle time by asking him or her to walk backward to the next activity.

3. Plan a unit on outer space and/or spaceships. Use cylinder shapes, such as paper-towel tubes, covered with aluminum foil and let the children create their own spaceships. The children can use stickers, construction paper, art foam, pipe cleaners, and other materials to design unique spaceships. Take the children outside and let them count backward as they prepare to blast off their spaceships.

Listening Activities

1. Randomly give each child a number (10–0) and ask him or her to count backward from that number down to 0.

2. Ask each child to count backward from 10 to 0.

Selected Literature

Counting Crocodiles by Judy Sierra (Harcourt, 2001)

The Counting Race by Margaret McNamara (Aladdin, 2003)

Five Little Dinosaurs by Will Grace (Scholastic, Inc., 2004)

Five Little Monkeys Sitting in a Tree by Eileen Christelow (Houghton Mifflin, 1999)

A Frog in the Bog by Karma Wilson (Simon & Schuster, 2003)

The Right Number of Elephants by Jeff Sheppard (HarperCollins, 1992)

Silly Sally by Audrey Wood (Harcourt, 1999)

Ten Little Fish by Audrey Wood (Scholastic, Inc., 2004)

Ten Little Mummies by Philip Yates (Viking Books, 2003)

Ten, Nine, Eight by Molly Bang (Greenwillow Books, 1996)

Ten Silly Dogs by Lisa Flather (Orchard Books, 1999)

Ten Terrible Dinosaurs by Paul Stickland (Puffin Books, 2000)

Home Letter

Hello,

We have planned a fun-filled activity for our next class meeting. We will be learning the math skill of counting backward. To make this skill more meaningful, we are planning a Backward Day at school. Could you please send your child to school on _____ wearing an article of clothing put on backward? We will begin our day with our closing circle and end the day with our greetings. Your child will be able to tell you all about our crazy activities at the end of the day.

If your child is hesitant to come to school with his or her clothes on backward, please do not be concerned. Very often when a child sees his or her friends dressed in this silly way, he or she will want to join in the fun. Even if your child does not participate, we are confident that a lot of learning will be going on at school.

Thank you for all of your assistance.

Sincerely,

_____

Measuring

Educational Objectives: The child will be able to identify some tools used for measuring such as a ruler, yard stick, measuring tape, scale, hour glass, thermometer, measuring cups, and measuring spoons. The child will be able to measure using both standard and nonstandard methods of measuring.

Vocabulary: *Measure*—the dimensions, capacity, etc., of anything as determined by a standard

Nonstandard measurement—not falling into the category of standard measurement tools (*Examples:* shoe, hand, block)

Standard measurement—something established for use as a rule or basis of comparison in measuring, such as inches, centimeters, cups, etc.

Classroom Activities

1. Before this circle activity, send home a letter (page 194) asking each child to bring a favorite toy to school to measure.

2. Introduce the concept of nonstandard units of measurement at circle time. Begin by trying to measure the teacher's favorite toy with a common item such as a piece of string. Talk about how these methods of measuring are not very reliable. Then read the book, *How Big Is a Foot?* by Rolf Myller. This book covers the concept of standard units of measurement. Introduce several different tools used for measuring such as a tape measure, ruler, and yardstick. Demonstrate how to measure with a ruler, introducing the term *inch* (*centimeter*). Make sure to stress that measuring starts at the *end* of the ruler closest to the number one. Let each child measure the toy brought from home and record the measurements on a chart. (Use all inch/centimeter measurements.) Let the children analyze who brought the longest toy and the shortest toy.

3. Before school the teacher should color the Inchworm Pet (page 196). Cut it out, and fold it accordion-style to make it look like it is moving in an inching motion. Attach the inchworm to the teacher's shoulder so it will be quickly noticed when the children arrive at school. Give a copy of the Inchworm Pet to each child and encourage him or her to make a pet to take home. Older children will be able to fold this creature accordion-style, but the teacher will need to assist the younger children. Attach the inchworms to the children's shoulders with tape.

4. Plan a cooking activity so that the children can use measuring tools found in the kitchen. A popular and fun activity is making pretzels. (See the recipe on page 194.) After the children have made the dough, encourage them to use it to roll out a shape, numeral, or letter.

5. Add assorted measuring tools to the Math Center. Include a balance scale and a bathroom scale, measuring cups, measuring spoons, and rulers. Add a large bin of dried beans, bottle caps, rice, etc., for the children to measure. Demonstrate how to use each tool properly. Set up water play on a different day with funnels and common volume measuring items.

6. Older children can be introduced to the concept of temperature. Purchase a large thermometer with large numerals. Place it outside the classroom where the children will have easy access to it. Cut out the Thermometer Pattern (page 197). Separate the dial from the rest of the paper. Glue the paper onto construction paper or tagboard and laminate it. Add the pointer of the thermometer with a brad. Explain that the lower numbers represent cooler weather and the higher numbers represent hotter temperatures. Point out the pictures on either side of the thermometer. Explain how the thermometer has big numerals advancing by 10s. Point out the small marks that represent the smaller numbers. Each day at circle time, check the real thermometer and have a child turn the dial on the paper model so it represents the same temperature. Record the temperature to the nearest 10 degrees. Always read the temperature at the same time each day.

Measuring *(cont.)*

Classroom Activities *(cont.)*

7. Measure each child to find out how many inches (centimeters) tall he or she is. Make a chart to show the shortest to the tallest (review of seriation) and label it with each child's name. (Omit this activity if you have a child who is exceptionally short or tall, as you do not want to make any particular child feel awkward.)

8. Older children can follow up with more measuring fun by doing this group activity. Provide each child with a ruler of his or her own. (Purchase rulers that only have inches [centimeters] represented.) Give each child a plastic, resealable bag with approximately 10 items to measure. (*Example:* paper clip, bobby pin, candle, piece of ribbon, craft stick, pipe cleaner, piece of tagboard, crayon, paintbrush) Try to get items that measure in 1" (2.54 cm) increments. Encourage the children to measure each item and tell you their results. Check often to make sure they are using the ruler correctly. As the children finish measuring the items in their bags, encourage them to find things in the classroom to measure. Encourage the children to take their special rulers home and measure things at home.

Listening Activities

1. Place assorted measuring items in a bag or mystery box. Let each child reach in and pull out one item and identify it.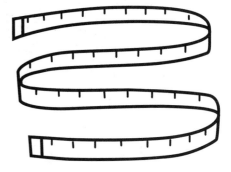

2. Collect an assortment of items to measure. Make sure they are all less then 12" (30 cm). Let each child select one item, identify it, and measure it using a ruler. Have the child share his or her results with the class.

3. Display a tray of various items. Ask a child to pick two items and compare their sizes. Encourage him or her to use vocabulary such as *longer, shorter, wider, thinner, heavier, lighter,* and *taller*.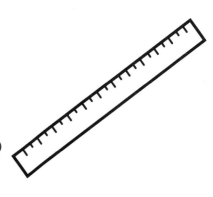

Selected Literature

Chickens on the Move by Pam Pollack (Kane Press, 2002)

Fannie in the Kitchen by Deborah Hopkinson (Aladdin, 2004)

How Big Is a Foot? by Rolf Myller (Sagebrush, 1999)

Inch by Inch by Leo Lionni (HarperTrophy, 1995)

Inchworm and a Half by Elinor Pinczes (Houghton Mifflin, 2003)

Jim and the Beanstalk by Raymond Briggs (Putnam, 1997)

Length by Henry Pluckrose (Scholastic Library Publishing, 1995)

Let's Measure It by Luella Connelly (Creative Teaching Press, 1996)

Me and the Measure of Things by Joan Sweeney (Dragonfly Books, 2002)

Measuring by Marcia Gresko (Gareth Stevens Publishing, 2004)

Home Letter and Recipe

Hello,

This month we will be learning about various measurement tools. We will be introducing our children to common measurement tools such as the ruler, yardstick, measuring tapes, measuring cups and spoons, scales, etc. These items will be available for your child to use during free play in the Math Center. To begin, we will be learning about the unit of measurement called the *inch*. Children will learn how to measure a favorite toy using a ruler, yardstick, or measuring tape. Please remind your child to bring his or her favorite toy to school on _____.

We will teach each child how to measure the toy correctly and record our data. It will be fun to see who has the longest and shortest toy.

Make sure to check the chart with the results from our day of measuring when you come to school.

Thank you for all of your assistance.

Sincerely,

_____

Use with Classroom Activity #4 on page 192.

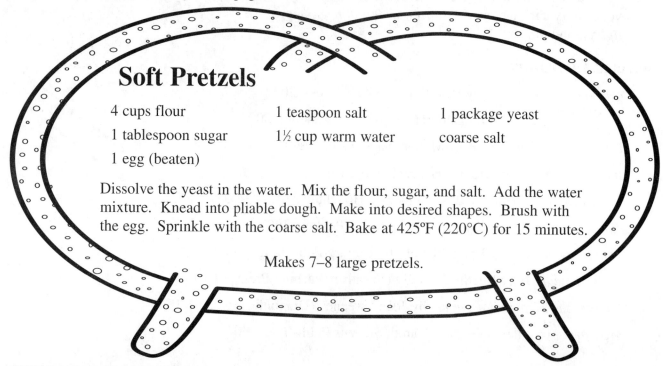

Soft Pretzels

4 cups flour	1 teaspoon salt	1 package yeast
1 tablespoon sugar	1½ cup warm water	coarse salt
1 egg (beaten)		

Dissolve the yeast in the water. Mix the flour, sugar, and salt. Add the water mixture. Knead into pliable dough. Make into desired shapes. Brush with the egg. Sprinkle with the coarse salt. Bake at 425°F (220°C) for 15 minutes.

Makes 7–8 large pretzels.

Beanbag Slide   **1–4 Players**

Developing Skill: Players will use a long measuring tape and practice measuring skills to see who can make the beanbag travel the farthest.

Materials

- 25' (7.5 m) measuring tape
- 4 different colored beanbags
- masking tape (on the floor to designate where to start)

Preparation: Unwind and lock the measuring tape and place it on the floor in an open space. Place a strip of masking tape on the floor at the beginning of the measuring tape to designate where to start the beanbag slides.

Playing the Game

1. The teacher demonstrates how to slide a beanbag across the floor, starting at the masking tape. Stress that the beanbag must stay on the floor as it slides so the players will be safe.

2. The player reads the number on the measuring tape after each slide to see how far it traveled.

3. Players try to improve their distance with each try.

Measuring Concentration   **1–4 Players**

Developing Skill: Players will verbally identify common tools used for measuring and find two matching pictures by remembering where the cards are on the table.

Materials

- Measuring Concentration Cards (pages 198–199)
- table

Preparation: Copy two sets of the cards. Cut them out and glue them onto construction paper. Laminate for durability.

Playing the Game

1. Place the cards on the table facedown.
2. Select who will go first. The player turns over two cards. If he or she finds a match, the player gets to keep the cards and take another turn. If the cards do not match, they are turned over and the next player gets to go.
3. As more cards are turned over, it gets easier to find a match.
4. Play continues until all the cards are matched up. The player with the most matching sets is the winner.

Use with Classroom Activity #3 on page 192.

Inchworm Pet

Use with Classroom Activity #6 on page 192.

Thermometer Pattern

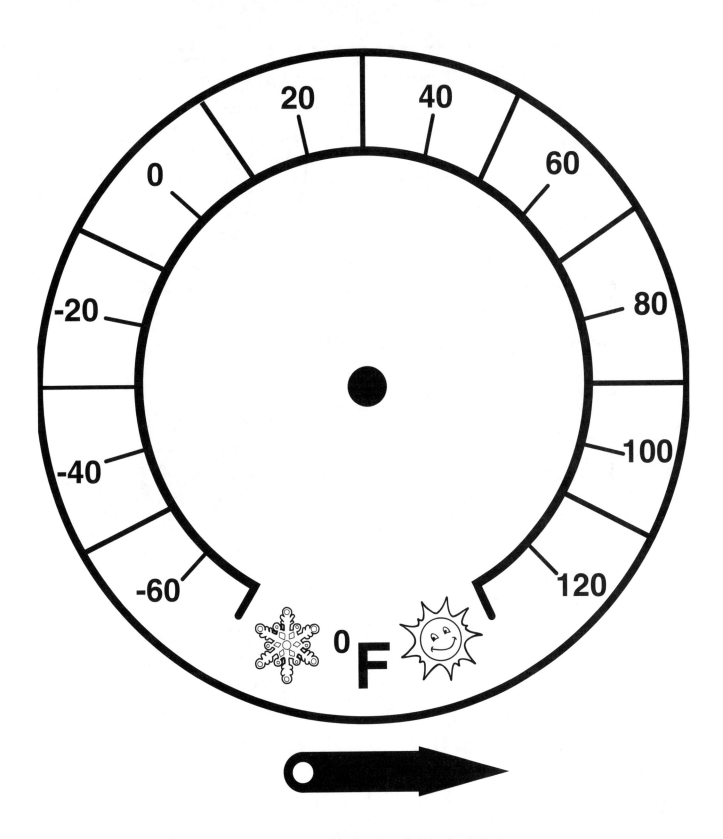

Measuring Concentration Cards

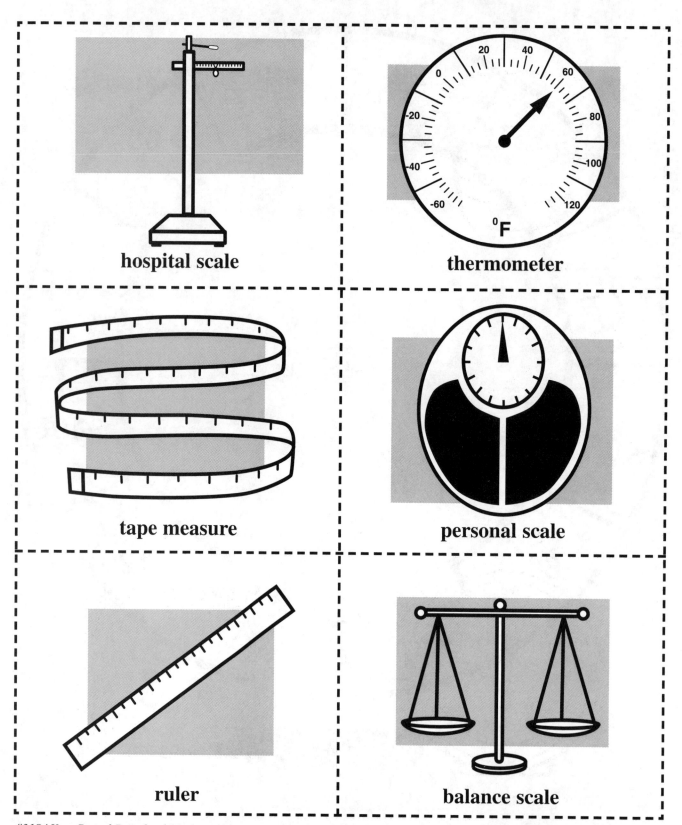

hospital scale

thermometer

tape measure

personal scale

ruler

balance scale

Use with the Measuring Concentration game on page 195.

Measuring Concentration Cards *(cont.)*

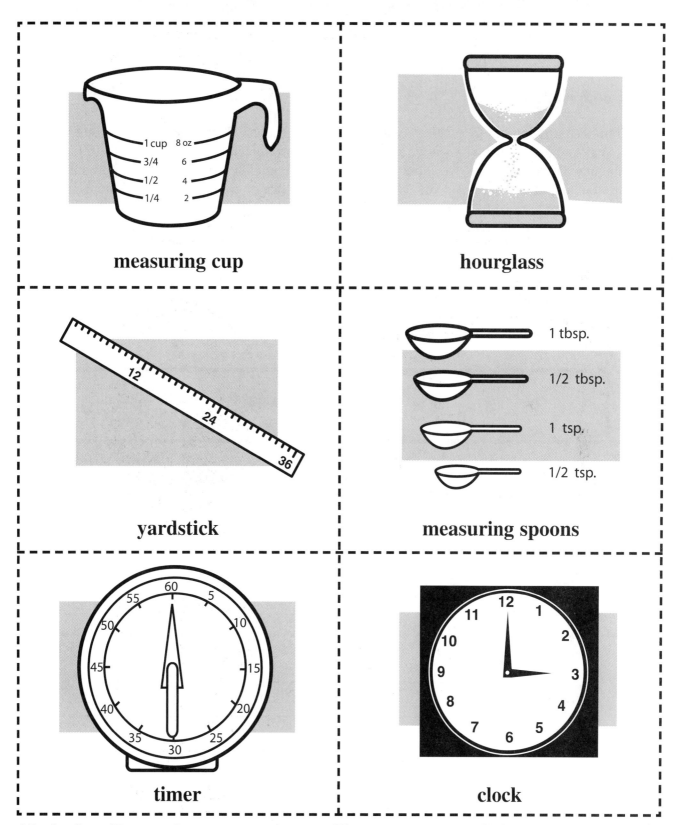

measuring cup

hourglass

yardstick

measuring spoons

timer

clock

Equal, More, and Less

Educational Objectives: The child will be able to accurately compare sets of equal, more (most), and less (least), and use the language of comparison.

Vocabulary: *Equal*—of the same quantity, size, number, value, degree, intensity, etc.

More—used to compare, meaning greater in number

Less—used to compare, meaning not as much, smaller, fewer

Classroom Activities

1. Introduce this concept by playing a game at circle time called Guess My Number. Use tagboard to make a number line with sliding end pieces. Use the Numeral Cards (pages 168–169). (See the diagram below.) Tack the sliding number line on a bulletin board. Tell the children that you are thinking of a number and you will use the number line to give them clues of "more" or "less" to help them figure out the number. Review the definitions of the words *more*, *less*, and *equal*. Ask one child to start the guessing. For example, select the number *6*. If the child guesses the number *3*, you can slide the bar on the left up to the 3 and say, "My number is more than 3." The next child may guess the number 8. This time, slide the bar on the right to the 8 and say, "My number is less than 8." Emphasize that the mystery number must be between the two sliding bars. Continue until it is obvious what the chosen number is. When the children get comfortable with the first number line, make a new one with higher numbers.

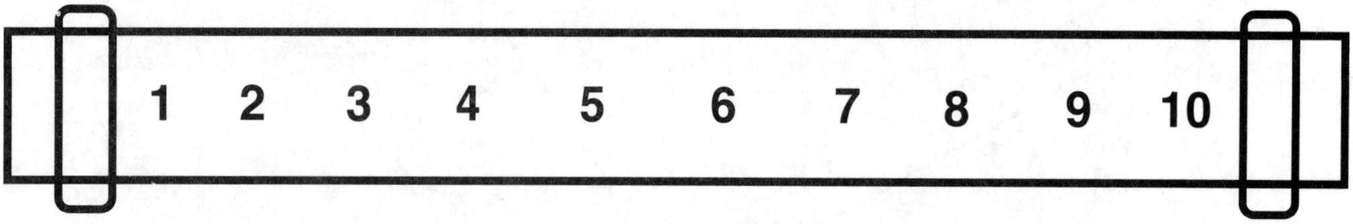

| 1 | 2 | 3 | 4 | 5 | 6 | 7 | 8 | 9 | 10 |

2. Have a Shape Hunt. Choose a particular shape or symbol that fits in with the theme you are studying. (*Examples:* triangle, heart, shamrock, flower, etc.) Make various colors or sizes of the shape and hide them while the children are out of the classroom. Have them hunt for the shapes as a group. After the hunt, sit with each child, count the shapes, and fill in his or her individual tally card (see below). Compare sizes and colors. Remember to include 0 where appropriate. This is a good activity to assess whether each child has grasped the concept of more, less, and equal. Let the children use the shapes and tally sheet to create a unique piece of art.

My _____ Hunt

I found the most _____.

I found the least _____.

I found an equal amount of _____.

Equal, More, and Less *(cont.)*

Classroom Activities *(cont.)*

3. Give each child a small fruit snack bag. Ask the children to sort the different types of fruit to find out which they have the most of, the least of, or an equal amount of. Use different terms, such as *more, fewer, same,* and *amount* to expand their vocabulary.

4. Challenge the older children's problem-solving abilities with some examples of *conservation.*

   - Line up 10 pennies with a space between each penny. Line up 10 more pennies so each penny is touching. (Do not count the pennies as you line them up!) Ask the child, "Which row of pennies has more?" Children who have mastered the concept of conservation will realize that if they count the two rows they will find the same (equal) number of pennies. Children who have not mastered this concept will say that the row of pennies with small spaces between the pennies has more because it takes up more room.

   - Continue with an example of conservation of volume. Show the children two jars with colored water. One jar is tall and thin (an olive jar works well); one jar is short and fat (e.g., a peanut butter jar). Fill the jars with the exact same amount of water before showing them to the children. Ask the children, "Which jar has more water?" After a class discussion, demonstrate that the two jars have the same (equal) amount of water by pouring the contents of each jar into measuring cups.

   - The last example of conservation can be demonstrated with two items made with play dough. Before the children arrive at circle time, measure out two similar amounts of play dough. Roll one into a ball and the other into a snake shape. Present the two shapes to the children and ask, "Which shape has more dough?" Again after a class discussion, demonstrate how the two items are made from the same (equal) amount of play dough by rolling the snake back into a ball shape. Children are usually very engaged, and this activity helps promote critical thinking.

Listening Activities

1. Use the balance scale from the Math Center. Place a few cubes on one side and a different number of cubes on the other side. Ask the child which side has more, less, etc. Change the number of cubes on the scale with each child.

2. Use a container of cubes or another fun manipulative. Randomly place a handful of cubes on the carpet in front of the children. Ask each child to reach for a handful of cubes to place in a separate pile. Ask the child to count the cubes and tell if he or she has more, less, or an equal amount of cubes as the teacher.

3. Have each child choose a number on which to stand on the floor number line (see Classroom Activity #2 on page 154). Then ask him or her to jump to a number that is one more or less, two more or less, etc. Ask the child to tell you what number he or she landed on.

Selected Literature

Just Enough Carrots by Stuart Murphy (HarperTrophy, 1997)

A More or Less Fish Story by Joanne Wylie (Scholastic Library Publishers, 1984)

More, Fewer, Less by Tana Hoban (Greenwillow Books, 1998)

More Than/Less Than Game   **2 Players**

Developing Skill: Players need to compare numbers that are more, less, and equal to other numbers.

Materials

• standard deck of cards

Preparation: Remove the face cards from the deck of cards.

Playing the Game

1. Divide the cards evenly between two players, placing them in two facedown piles.

2. Each player flips over the top card on his or her pile.

3. The player with the higher card (more than) gets to keep both cards.

4. If there is a tie (equal), each player turns over another card. Then the player with the higher number gets all the cards again. Play continues until all the cards have been turned over.

5. The winner is the player with the most cards.

The Dinosaur Game   **2 Players**

Developing Skill: Players will read the words *more* or *less* on a spinner and count their dinosaurs to see who will win each round of the game.

Materials

• 25–30 small plastic dinosaurs

• 1 plastic cage (basket) large enough to hold all of the dinosaurs

• 1 standard die (with numerals 1–6)

• 1 spinner divided into 2 equal sections

Preparation: On the spinner, write *more* on one side, and *less* on the other side.

Playing the Game

1. Each player rolls the die and collects the corresponding number of dinosaurs from the basket. If the second child rolls the same number as the other player, he or she rolls again.

2. The two players then take turns spinning the More or Less spinner. If it lands on more, the player with more dinosaurs gets to keep all the dinosaurs from that round. If it lands on less, the player with fewer dinosaurs gets to keep all the dinosaurs.

3. Play continues until all the dinosaurs are out of the basket. The player with the most dinosaurs at the end of the game is the winner. (Toward the end of the game, if a player rolls a 6 and there are only 2 dinosaurs left, that is all that he or she will take.)

Money

Educational Objectives: The child will be able to identify the penny, nickel, dime, and quarter and recognize that coins have different values.

Vocabulary: *Penny*—a coin worth one cent   *Dime*—a coin worth ten cents
Nickel—a coin worth five cents   *Quarter*—a coin worth twenty-five cents

Classroom Activities

Note: Do not introduce all the coins at the same time; this is too confusing for the children. Instead, introduce one coin at a time and concentrate on it exclusively until the children have a solid grasp of that coin. Then introduce the next coin.

1. Use the coin patterns (pages 207–210) to make a large poster or individual posters, depending on what will work better for your children. Display them in the Math Center.

2. Introduce the copper penny at circle time. Describe the pictures on both sides of the coin. Explain the value of one cent. Draw the symbol for cents. Demonstrate by counting that five pennies equal five cents. Show how to write five cents. Show other examples (seven cents, nine cents, etc.) to help make this clear. Make up a poem, chant, or song to help the children remember that a penny is worth one cent. (*Example:* "A penny is worth one cent, one cent, one cent. A penny is worth one cent, I learned it at my school.") Demonstrate how to polish pennies with lemon juice and toothbrushes. Let the children continue this activity during free play.

3. Coin collection books are good visuals to use. Provide the children with magnifying glasses so they can observe the finer points of each coin. Challenge the children to find the year of their birthday on a coin. Explain that coins are made in Denver or Philadelphia and that a letter is stamped on each coin to show where it was made. Have the children sort the coins into two separate piles, depending on which letter is stamped on the coin—a "D" or a "P."

4. When the children are ready to learn about the nickel, review facts about the nickel. It is fun to introduce counting by 5s with the "Counting by Five's hands" (page 211 and 212). Copy the page twice—the first page of hands on one color and the second page of hands on another color. Glue the hands onto tagboard or file folders and laminate. Punch a hole in the wrist part and connect the handprints alternating the two colors. Use a removable metal ring to put the hands together. (All the numbers ending with 0 should be one color and the numbers ending in 5 should be the other color.) The children can count by 5s by flipping through the hands. Demonstrate this activity and ask the children if they recognize a pattern. Counting by 5s is a form of an AB pattern with one number ending in a 5 and the other number ending in a 0. Later they can use this same tool to count by 10s.

5. Coin connectors can be purchased at most teacher supply stores. Introduce a Coin Connectors manipulative activity and add it to the Math Center. Demonstrate how the children can use pennies, nickels, and dimes to create designs.

6. Recycling aluminum cans is a good way for the class to work together to earn money. Check to see if there is a recycling center near the school. Send home a letter to parents asking them to help save, clean, and crush cans for a period of one month. Select a date for all the cans to be brought to school in plastic bags. Take the class on a field trip to see how the cans are sorted, weighed, and crushed to be shipped to a factory. (If the class cannot go on a field trip, arrange for someone to take the cans to the recycling center.) Purchase something for the classroom with the money earned.

Money *(cont.)*

Classroom Activities *(cont.)*

7. Have the children "buy a lunch ticket" before eating lunch. (Choose a price for lunch that the children will be able to handle realistically. It could be something as low as five cents.) Provide a tray with several different coins. Let each child come up and count out the correct amount of money to purchase the ticket. Copy fun tickets to hand out. This activity can also be used to buy tickets to the playground, etc.

8. At the end of the year, have a raffle of memorable classroom items. Make copies of raffle tickets with a specific price that everyone in the class has mastered. Provide an ample supply of coins as well. It is helpful to place the money on a tray so it can be spread out and is easy to see. Review all the items in the raffle. Make sure you have more items than the number of children in the class so the last few still have an assortment of things to choose from. Let each child come up and buy one ticket using the coins on the tray. Have each child write his or her name on the ticket. Put the tickets in a bowl and call one name at a time. Each child can purchase one classroom item. (Suggestions for raffle items include class-made books, posters, plants, charts, leftover play dough, silly putty, etc.)

Listening Activities

1. Place pennies on a tray. Ask each child to count out a specific number of cents.

2. Place pennies and nickels on a tray. Ask each child to select two to four coins and identify how many of each coin he or she has selected. (Add the dime and quarter as they are introduced.)

3. Prepare a shopping bag filled with small toys with price tags attached to them. Make sure the prices are at a low enough level that the children can succeed in this game. Include a collection of money. Let each child reach into the bag and pull out a toy. Have him or her identify the cost of the toy and then count out the proper amount of money needed to pretend to buy the toy. (This activity can be done with small party favors. Children can keep what they purchase.)

Selected Literature

All About Money by Erin Roberson (Children's Press, 2004)

Bank in a Book by Jim Talbot (Innovative Kids, 1999)

The Big Buck Adventure by Shelley Gill (Charlesbridge Publishing, 2002)

The Coin Counting Book by Rozanne Williams (Charlesbridge Publishing, 2001)

Counting Money by Tanya Thayer (Lerner Publishing Group, 2002)

Deena's Lucky Penny by Barbara Derubertis (Kane Press, 2004)

Learning About Coins by Rozanne Williams (Gareth Stevens Publishing, 2004)

Money Matters series by Mary Hill (Scholastic, 2005)

Pennies by Suzanne Lieurance (Children's Press, 2003)

A Quarter from the Tooth Fairy by Caren Holtzman (Scholastic, Inc., 1995)

Penny Coin Toss

2 Players

Developing Skill: Players will identify the difference between *heads* and *tails* on a coin.

Materials

- 1 penny
- 2 crayons
- 2 Penny Coin Toss Tally Sheets (page 213)

Playing the Game

1. One player chooses heads and the other player chooses tails.

2. Each player takes a tally sheet and a crayon and writes his or her name on the paper.

3. The players take turns tossing the coin.

4. If it lands on heads, the player who chose heads marks an "X" on one of the *heads* pictures. If it lands on tails, the other player marks an "X" on one of the *tails* pictures.

5. The players continue taking turns tossing the coin until one player has crossed out all 10 pictures on his or her tally card.

Nickel Game

2 Players

Developing Skill: Players will read the number on the gameboard and count the corresponding number of pennies and put them in his or her wallet.

Materials

- 1 Nickel Game Gameboard (page 214)
- 1 cardboard box
- 1 table
- 2 small wallets
- 20 pennies
- 1 nickel

Preparation

Glue a copy of the gameboard onto the lid of the box. Place the pennies in a pile on the table.

Playing the Game

1. Select who will go first. The first player holds the nickel on top of his or her fist approximately 8" (20 cm) above the gameboard and drops it. The object is to get the nickel to drop on a large number. The player will count out the corresponding number of pennies and put them in his or her wallet.

2. Play rotates between the two players.

3. When all the pennies have been collected, the players count the money in their wallets. The player with the most money is the winner.

Toy Shopping Concentration

2–4 Players

Developing Skill: Players must be able to match the money value on the toy price tag with the money value on the coin card.

Materials

- Coin and Toy Cards (pages 215–216)

Preparation: Copy the Coin and Toy cards. Cut out the cards and glue them onto construction paper. Laminate for durability.

Playing the Game

1. Place the cards in rows facedown on the table.

2. Select who will go first. The player turns over two cards. The players must match a card with a specific money amount with a toy with the same purchase price. If he or she finds a match, then the player gets to keep the cards and take another turn. If they do not match, the cards are turned over and the next player gets to go.

3. Play continues until all the cards are matched up. The player with the most matching sets is the winner.

Coin Lotto

2–4 Players

Developing Skill: Players match the money amount on the game cards with matching money amounts on their gameboard.

Materials

- Coin Lotto Gameboard (page 217)

Preparation: Make eight copies of the Coin Lotto Gameboard. Lightly color the penny to help identify that it is made of copper. Glue four of the gameboards to different colors of construction paper. Glue the other four gameboards to one complementary color. Laminate for durability. Cut apart all of the same-colored gameboards to make individual game cards. Stack the game cards.

Playing the Game

1. Give each player a gameboard.

2. Select who will go first. This player turns over the top card. If the card matches a money amount on his or her gameboard, it is placed on top of the picture on the gameboard. If that picture is already covered, it is placed faceup in a discard pile.

3. The next player can pick from either pile.

4. Play continues until someone has a full gameboard.

5. When the teacher is modeling how to play this game, it is important to use money vocabulary. (*Example:* "I have two dimes. That would be 20 cents.")

Use with Classroom Activity #1 on page 203.

Penny Patterns

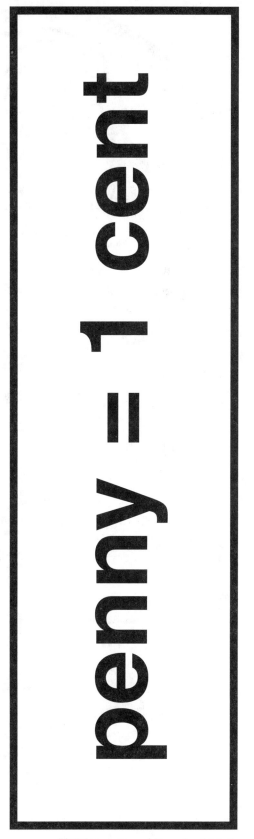

Nickel Patterns

nickel = 5 cents

Use with Classroom Activity #1 on page 203.

Dime Patterns

dime = 10 cents

Use with Classroom Activity #1 on page 203.

Quarter Patterns

quarter = 25 cents

Use with Classroom Activity #4 on page 203.

Counting by Fives

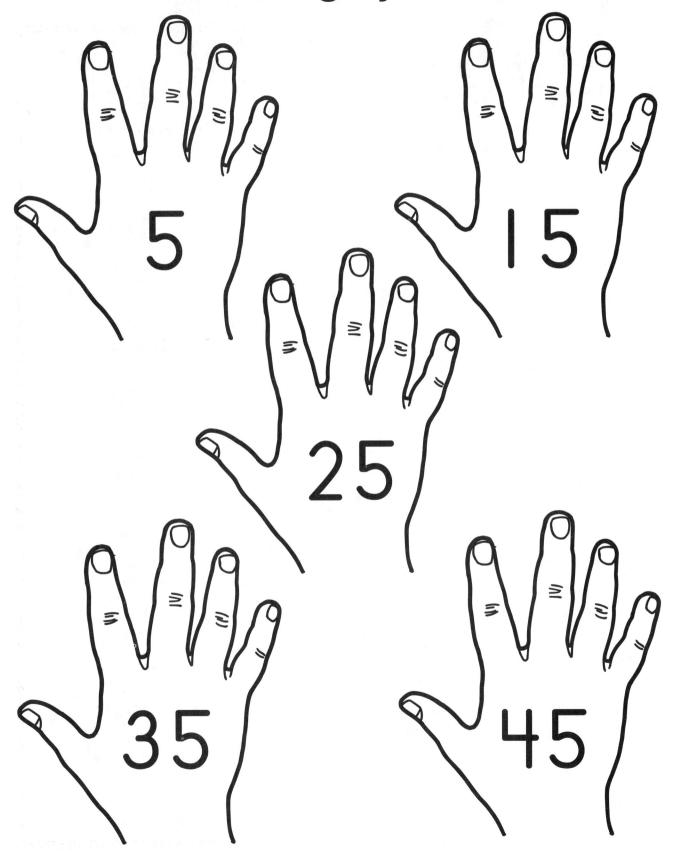

Counting by Tens

Use with the Penny Coin Toss game on page 205.

Penny Coin Toss Tally Sheet

Name:_____

Heads		Tails	

Use with the Nickel Game on page 205. Color each section a different color before laminating.

Nickel Game Gameboard

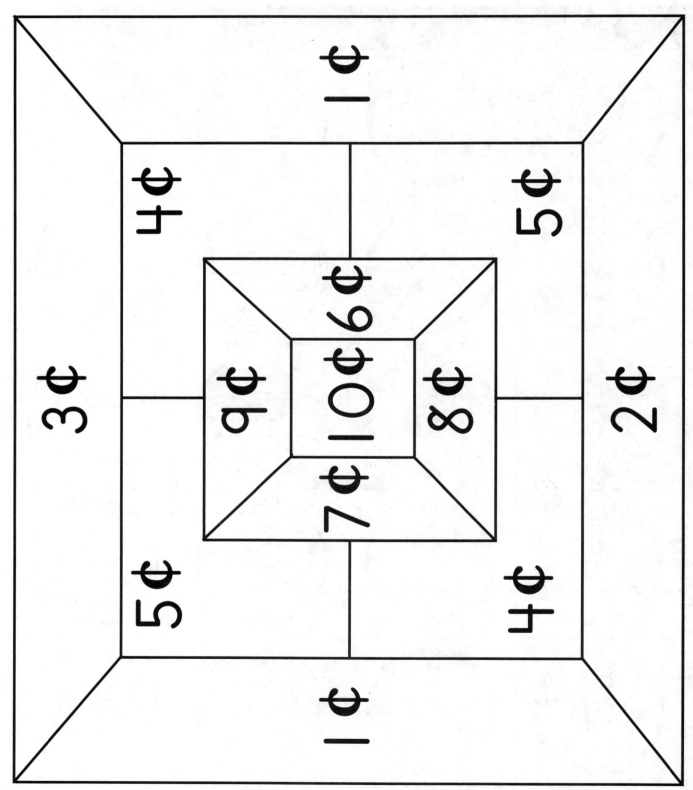

Use with the Toy Shopping Concentration game on page 206.

Toy Shopping Coin Cards

10¢	1¢	25¢
9¢	7¢	6¢
8¢	3¢	15¢
4¢	2¢	5¢

Use with the Toy Shopping Concentration game on page 206.

Toy Shopping Toy Cards

Use with the Coin Lotto game on page 206.

Gameboard

Coin Lotto

20¢	11¢	26¢
5¢	1¢	2¢
6¢	10¢	25¢

Area Code and Telephone Number

Educational Objective: The child will be able to independently recite his or her telephone number, including the area code.

Vocabulary: *Telephone Number*—the specific seven numbers assigned to a personal residence telephone or cell phone

Area Code—any three numbers assigned as a telephone code to each of the areas into which the United States and Canada are divided

Classroom Activities

1. Introduce telephones by reading a book about phones. Tell the children they will be able to make a pretend cell phone to take home. Use ¼" (.6 cm) foam board. Cut the foam board to approximately 3" x 6" (8 cm x 15 cm). Use color-coded dots to represent the 12 buttons on the phone. (These can be purchased at any office supply store.) Demonstrate how to place the dots on the pretend phone to match a real telephone. Explain what the * and # buttons are used for on a telephone. Add small label stickers to simulate the ear piece and speaker. Have the child add small dots. Gently push a pipe cleaner into the top of the phone to simulate an antenna. Put a larger label on the back with the child's name and home number.

2. Look in a large local telephone book to find a map showing the different area codes in the state. Make a copy of it. Highlight the area code for the school's location. Enlarge the map and color the different areas codes using different colors. Use this as a visual example of the different area codes surrounding the school.

3. Make up a poem, chant, or song to help the children learn the local area code. Have the children work together to make a large wall poster of the area-code poem. (See sample below.) Write the lines of the chosen poem on bulletin board paper. Then add very large numerals. Do not color in the numbers; instead, let the children tear paper into 1" (2.54 cm) pieces. Glue inside the lines on the numerals to make a large mosaic representation of the area code. Display the poster where the children can see it and use it often to review this important information.

4. Add two real (but not plugged in) telephones to the Math Center. Make a copy of the Telephone List (page 220). Laminate it and place it by the phones. Encourage the children to practice calling their home telephone number, as well as the different numbers on the Telephone List. Make copies of the Telephone List for each child to take home. If appropriate, let each child select 2–3 friends' names to add to the bottom of the list.

To make sure I'm safe

It's good to know

My area code is

Area Code and Telephone Number *(cont.)*

Classroom Activities *(cont.)*

5. In some areas a telephone number is available to get a local weather report. If this is offered, let each child try dialing the number independently to listen to the weather report. In addition, some local libraries have a "Dial-a-Story" program. If this is available in your area, send home a letter to the parents letting them know about this service and provide the local phone number.

6. Review the 911 emergency number if it is available in the area surrounding the school. Emphasize that it is only to be used for emergencies. Review examples of real emergencies and discuss when it would not be appropriate to call 911.

7. Make a giant telephone keyboard so the children can practice jumping on the numbers of their telephone numbers. Use craft foam and cut out twelve 8" (20 cm) squares. Use a permanent marker to add the numbers and symbols on a phone. Use masking tape to apply the squares to the floor. Let the children recite their telephone numbers as they jump on the corresponding numbers.

Listening Activities

1. Ask each child to recite his or her personal telephone number, including the area code.

2. Use the foam phone number squares on the floor. Let each child recite and jump out his or her phone number.

3. Prepare a list of examples of when it would be good to call 911. Include some obvious examples of when it would not be good to call 911. Read the list and ask each child, "Should I call 911?" "Why or why not?"

4. Ask each child to recite the emergency number.

Selected Literature

Five Two, Five Blue by Molly Griffis (Eakin Press, 1999)

My First Phone Call by Julia Allen (Aro Books Inc, 1997)

My Talking Telephone by Johnny Miglis (Barnes and Noble Books, 2003)

Rosie's Surprise (Fairy Phones) by Louise Comfort (Barron's Educational Series, 2003)

Toni's Topsy-Turvy Telephone Day by Laura Ljungkvist (Harry N. Abrams, Inc. 2001)

Telephone List

911	**Emergency**	_____
★	**Police**	_____
🚑	**Ambulance**	_____
🔥	**Fire Department**	_____
🍕	**Pizza Shop**	_____
🎥	**Movie Theater**	_____
🐱	**Vet Office**	_____
🍞	**Bakery**	_____
👦	**Friends**	_____ _____

Addition

Educational Objectives: The child will be able to add two numbers (0–5) together to get a new number called a *sum*. He or she will be able to identify the plus sign, the equal sign, and recognize a written equation.

Vocabulary: *Add*—to combine numbers into a sum

Plus sign—a symbol indicating addition

Equal sign—a symbol indicating the same quantity, size, value, number, etc.

Equation—a statement of equality shown by the equal sign

Classroom Activities

1. Demonstrate a simple addition problem using items from the classroom such as blocks, cars, etc. Next, introduce the plus and equal signs using the large patterns on page 224. Let the students get involved. Use hula hoops to help demonstrate an addition problem. Place two small hoops on either side of a plus sign. Working from left to right, add an equal sign and then a larger hoop. Call up different numbers of children to stand in the hoops. Ask the children what they think the sum will be. Check the answer by asking the children to move to the larger hoop. Count out the answer or have the children do a count off to find the answer. Do this enough times so everyone will have a special turn to stand in the hoop.

2. Write an equation on a dry-erase board or chalkboard going from left to right. Provide manipulatives to demonstrate the equation. Show the same equation written in vertical format. Explain that they are both equations used to show an addition problem.

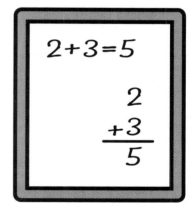

$$2+3=5$$

$$\begin{array}{r} 2 \\ +3 \\ \hline 5 \end{array}$$

3. Make a poster of each child's first and last name. Put blanks for a math equation below the name. Ask each child to count the letters in his or her first name. Put this number under the first name. Count the letters in his or her last name. Put this number under the second name. Help the child do the math equation to find out how many letters are in his or her whole name. Mount the poster on colorful paper. Let the child take this home and place it where it can be observed often. This will help the child learn addition skills, as well as his or her last name.

4. Encourage your children to make up math equations of their own using the supplies in the Math Center. Use connecting cubes with laminated copies of the plus and equal signs (page 224). Cut out the interiors of plastic coffee can lids to make mini hula hoops. Model how to do this activity before adding it to the Math Center.

5. Use vertical addition cards in a Flip Box (see directions on page 162) in the Math Center.

6. Introduce the Jolly Bunnies and Jelly Beans take-home activity (page 225) at circle time. Explain that all of the children will get a small bag of jelly beans to take home to do addition problems with their parents. Send home a parent letter (page 223) explaining the activity along with the activity sheet and beans. (This learning experience can be sent home with all of the children at one time, or can be spread out over a longer period of time with only a few children receiving the project at a time.) Encourage the children to share their math papers with the class when they bring them back to school.

Addition *(cont.)*

Listening Activities

1. Choose a number such as 5. Tell the children all the addition problems you will present will have a sum of 5. The teacher puts two hands behind his or her back, brings out one hand, and reveals a random number of manipulatives (0–5). Ask the child how many would need to be in the other hand to add up to 5. Bring out the other hand to show if he or she is correct. Repeat this activity for each child, alternating problems.

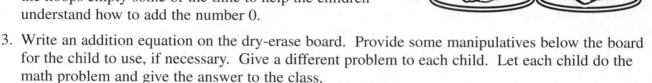

2. Use the same hula hoops used for Classroom Activity #1. Call different children to stand in the hoops and let one child at a time solve a math problem. Remember to keep one of the hoops empty some of the time to help the children understand how to add the number 0.

3. Write an addition equation on the dry-erase board. Provide some manipulatives below the board for the child to use, if necessary. Give a different problem to each child. Let each child do the math problem and give the answer to the class.

Selected Literature

Adding by Rozanne Williams (Gareth Stevens Publications, 2004)

Adding It Up at the Zoo by Judy Nayer (Yellow Umbrella Books, 2002)

Addition Annie by David Gisler (Children's Press, 2002)

Animals on Board by Stuart Murphy (Harper Trophy, 1998)

Building Numbers by Jenny Fry (Barron's Educational Series, 2002)

Domino Addition by Lynnette Long, Ph.D. (Charlesbridge Publishing, 1996)

Hershey's Kisses Addition Book by Jerry Pallotta (Scholastic, Inc., 2001)

Little Number Stories: Addition by Rozanne Williams (Creative Teaching Press, 1995)

Little Quack by Lauren Thompson (Simon & Schuster Children's Publishing, 2003)

Monster Math Picnic by Grace Maccarone (Scholastic, Inc., 1998)

More Bugs, Less Bugs by Don Curry (Capstone Press, 2000)

Oliver's Party by Jenny Fry (Barron's Educational Series, 2002)

Quack and Count by Keith Baker (Harcourt, 1999)

Take Two Pirates by Tim Healey (Readers Digest, 2003)

Dominos Addition   **2 Players**

Developing Skill: Players need to add the dots on both sides of the domino to find out the sum.

Materials

- set of dominos with dots (1–6)

Playing the Game

1. Divide the dominos in half between two players.

2. Each player turns over one domino and adds the dots on each side of the domino.

3. The player with the highest sum takes both dominos.

4. If the sum is the same for both players, they turn over two new dominos and complete a new problem. Whoever has the highest sum takes the dominos from both plays.

5. Play continues until all the dominos have been played.

6. Each player counts his or her dominos. Whoever has the most is the winner.

Hello,

We are sending home an addition activity for you to do with your child. We have been learning about addition, and the children should be familiar with the steps needed to complete a simple problem. Your child should first write his or her name at the top of the paper with a crayon or marker.

Have your mathematician place a random number of beans on the first plate and another group of beans on the second plate. Next, have your child add to find out how many beans there are all together. He or she should be able to show you how to push all the beans together onto the larger plate at the end of the equation to check the *sum* (answer). Encourage your child to do several different problems. Take turns giving problems to each other. When you are ready to finish, have your child illustrate a problem by coloring beans on the plates. Then assist him or her in writing the numbers on the blanks under each plate. Enjoy the jelly beans.

Make sure the paper comes back to school tomorrow. We will share the pictures and display them for all to enjoy. Have fun and thank you for you cooperation.

Sincerely,

_____

Addition and Equal Sign Patterns

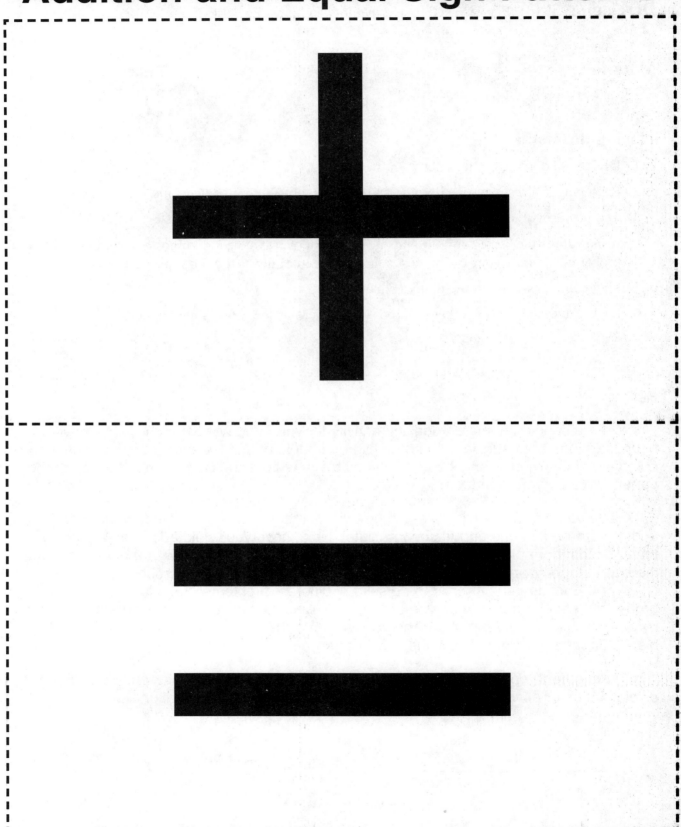

Jolly Bunnies and Jelly Beans

Name

Subtraction

Educational Objectives: Given two numbers between 0–10, the child will be able to solve a subtraction problem by taking one number away from another number to find the difference. He or she will be able to identify a minus sign and recognize a subtraction equation.

Vocabulary: *Minus sign*—a symbol indicating subtraction

Equation—a statement of equality shown by the equal sign

Difference—the amount left after subtraction

Classroom Activities

1. Introduce subtraction by adding the minus sign to the symbols that have been used in class. (See page 228.) Use manipulatives to demonstrate a simple subtraction problem.

2. Pass out small plates and ten small food items to each child at circle time. Explain to the children that they are going to learn about subtraction with the food in front of them. Use the dry-erase board and start with the equation $10 - 1 =$ _____. Let the children eat one food item and ask them how many they have left. Write the completed equation on the dry-erase board. Proceed with different equations until all the food is gone. Don't forget to include an equation with 0.

3. Add subtraction cards to the Flip Box (see page 162) in the Math Center.

Listening Activities

1. Set up subtraction problems using small manipulatives. Ask each child to take a turn being your assistant. Verbalize a problem and let the assistant remove the number of manipulatives that you stated in the problem. Then ask him or her to give the answer to the problem by repeating the complete equation.

2. Draw a subtraction equation on the dry-erase board or chalkboard. Provide manipulatives in front of the board. Ask each child to use the manipulatives to solve the problem and then tell the completed equation to the class. Give a new problem to each child. Remember to include equations with 0.

Selected Literature

Building Numbers by Jenny Fry (Barron's Educational Series, 2002)

The Doorbell Rang by Pat Hutchins (HarperTrophy, 1994)

Elevator Magic by Stuart Murphy (HarperTrophy, 1997)

Hershey's Kisses Subtraction Book by Jerry Pallotta (Scholastic, Inc., 2002)

Little Number Stories: Subtraction by Rozanne Williams (Creative Teaching Press, 1995)

Monster Musical Chairs by Stuart Murphy (HarperTrophy, 2000)

Subtraction Fun by Betsy Franco (Capstone Press, 2002)

Take Away by Lisa Trumbauer (Capstone Press, 2003)

Ten Terrible Dinosaurs by Paul Stickland (Puffin Books, 2000)

Toy Box Subtraction by Jill Fuller (Children's Press, 2004)

Twenty is Too Many by Kate Duke (Dutton Books, 2000)

Musical Chairs   **Circle Time Game**

Developing Skill: Players will need to walk around chairs while the music is playing. They will need to quickly find a chair when the music stops. Players will need to remove themselves from the game if they cannot find a seat when the music stops.

Materials

- chairs (1 for each child)

- music that can be controlled by the teacher (record player, tape player, etc.)

Playing the Game

1. Set the chairs up back to back. Make sure there is enough room so the children can walk around the chairs.

2. Explain to the players that when the music starts, they are to walk around the chairs. When the music stops, they are to sit in the chair closest to them. Each time the music starts, one chair will be taken away, or subtracted, from the group of chairs. Each time the music stops, the player who is left without a chair will be subtracted from the game.

3. Play continues until only one player is left sitting on a chair.

Mouse and Cheese Game   **2 Players**

Developing Skill: Players will need to read a simple subtraction equation and find the answer. Players will need to know which child had the lowest difference to decide who will keep the cards.

Materials

- Mouse and Cheese Game Cards (pages 229–230)

Preparation: Add some color to the mice and cheese and glue the cards onto construction paper. Laminate them for durability and cut them apart.

Playing the Game

1. Divide the cards in half.

2. Both players turn over one card. Each reads his or her subtraction problem and announces the answer.

3. Whoever has the lower difference gets to keep the cards.

4. The goal is to avoid a mouse card. If a player turns one over, then the other player gets to keep the cards from that round.

5. If both players turn over cards with the same difference, then both players turn over two new cards on top of the first ones. The player with the lowest difference gets to keep the cards.

6. The player with the most cards is the winner.

Minus and Equal Sign Patterns

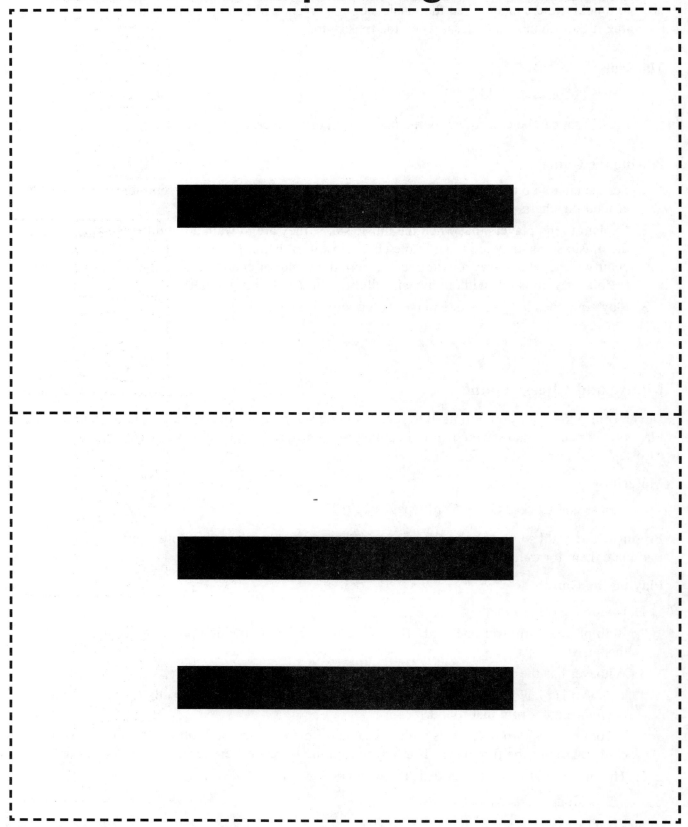

Use with the Mouse and Cheese game on page 227.

Mouse and Cheese Game Cards

Mouse and Cheese Game Cards *(cont.)*

Odd and Even

Educational Objectives: Given a number between 0–10, the child will be able to tell if it is an odd or even number. The child can have the assistance of any small manipulative.

Vocabulary: *Odd*—numbers that have a remainder of one when divided by two

Even—numbers that can be divided exactly by two

Classroom Activities

1. A good way to introduce this activity is to use small manipulative people that come in play sets. Line up two people and point out that they are partners. Two is an *even* number. Add another person and point that he or she is an *odd* number because he or she does not have a partner. Three is an odd number. Add another person and emphasize that this person is no longer odd because he or she has a partner. The number four is even. Continue this demonstration up to the number 10.

2. Review the chants learned in the Pairs section (see Classroom Activity #7 on page 179). *"Two, four, six, eight. Who do we appreciate?* Name of child or teacher, repeat, yea!" and *"One, three, five, seven, (pause) and nine. Who do we think is mighty fine?* Name of child or teacher, repeat, yea!" Explain that these cheers are a good way to remember numbers that are odd or even.

3. Teach the old-fashioned Odd/Even Hand Activity. Select a child to help demonstrate this activity to the class. Let the child choose "odds" or "evens." The teacher and child put one hand behind their backs. The teacher says, "One, two, three." At the word *three*, both the teacher and child simultaneously show their hand with either one or two fingers showing or a closed fist to represent the number zero. If the sum of the two hands is odd, the one who chose odds is the winner. If the sum is even, then the one who chose evens is the winner. (If both children show a fist, then they must do the activity again as zero is neither an odd nor an even number.) The only possibilities are the numbers 1, 2, 3, and 4, so this helps the children learn that 2 and 4 are even numbers and 1 and 3 are odd numbers. This activity is usually used to decide who will go first in a game or competition.

4. Let the children help make a large poster titled, "Odd and Even Numbers." Divide the poster into two columns and let the children look through magazines and newspapers to find numbers to cut out and glue onto the odd or even side of the poster.

Listening Activities

1. Pass out a number from 0–10 to each child. Call on one child at a time and ask if the number he or she is holding is odd or even. Have some manipulatives available so the child can arrange a corresponding number of items in groups of twos to see if the number is odd or even.

2. Have each child jump on the odd numbers on the number line, calling out the number as he or she lands. Next, have the child jump on the even numbers. This is a review of counting by twos.

Selected Literature

Bears Odd, Bears Even by Harriet Ziefert (Puffin Books, 1997)

Even Steven and Odd Todd by Kathryn Cristaldi (Scholastic, Inc., 1996)

Gray Rabbit's Odd One Out by Alan Baker (Houghton Mifflin, 1999)

Odd and Even Numbers by Shirley Tucker (Capstone Press, 2002)

Odd One Out by Jenny Tyler (Usborne Books, 2003)

Halves

Educational Goals: The child will be able to recognize that a whole item can be divided into two equal parts. The child will be able to physically divide a simple item into two equal halves.

Vocabulary: *Whole*—containing all the elements or parts; entire; complete

Half—either of the two equal parts of something

Classroom Activities

1. Introduce the concept of halves using play dough. Make a shape and use a plastic knife to divide it in half. Point out that both halves are equal. Make other shapes and divide each into two parts that are not equal. Explain that even though a shape is divided into two parts, it is not divided in half unless the parts are equal.

2. Read the book, *The Little Mouse, The Red Ripe Strawberry, and the Big Hungry Bear* by Audrey Wood. Copy the strawberry shape (page 234) onto red or pink paper. Make a copy for each child and cut it out. Label the strawberry halves before the children arrive at school.

   Cut the strawberries in half. Label the back of one half with an uppercase letter and the back of the other half with a matching lowercase letter. Place the halves in two separate piles. Each child will choose one uppercase-letter half and one lowercase-letter half to turn over and decorate. Make sure the children put their names on each half of the strawberry next to the alphabet letter. Decorate the berry by tearing small pieces of red or yellow paper and gluing them onto the two halves to simulate the small seeds on the berry. Add a green leaf on each half. When all of the halves have been completed, the child keeps the half with the upper case letter. The half with the lower case letter is placed on a table. Each child finds the lower case half that will match (like a puzzle) to make his or her strawberry whole again. Glue the two halves onto another piece of paper and label it with the two artists' names. Each child will have a completed picture made with a classmate to take home.

3. Let students paint with halves of fruits or vegetables. Apples, green peppers, bananas, star fruits, and potatoes all work nicely for this project. Before the activity, let each child cut one piece of fruit in half using a plastic knife, with the supervision of the teacher.

4. Teach the children to pour half a cup of beverage at snack time. Also provide a soft snack and with adult supervision, let the children cut it in half with a plastic knife.

5. Some companies sell assorted plastic fruits that are divided in half and then can be reconnected with Velcro®. Put these in a plastic bowl with plastic knives and add them to the Math Center. Encourage the children to have fun cutting the pretend fruit in half.

6. Encourage the children to make shapes with play dough and then cut them in half using plastic knives.

7. Let the children use 1 cup (225 g) and ½ cup (100 g) measuring cups with assorted dry media. Set up a large bin with beans, rice, cereals, assorted pastas, etc. Provide large spoons or scoops and enough measuring cups for the children playing in the area. Demonstrate how they can check the accuracy of the measuring cups by filing in the ½ cup two times to see if it fills in the 1 cup measurement.

Halves *(cont.)*

Listening Activity

Use a dry-erase board or chalkboard. Draw a different shape for each child. (*Examples:* square, circle, heart, oval, etc.) Draw a line on the shape that obviously does not divide the shape in half. Ask the child, "Did I divide this shape in half?" When the child answers, "No," ask him or her, "Why?" Then, ask the child if he or she can erase your line and correctly divide the shape in half to fix your error.

Selected Literature

Give Me Half! by Stuart Murphy (HarperTrophy, 1996)

Inchworm and a Half by Elinor Pinczes (Houghton Mifflin, 2003)

I Saw a Bullfrog by Ellen Stern (Random House Books for Young Readers, 2003)

Let's Fly a Kite by Stuart Murphy (HarperTrophy, 2000)

The Little Mouse, The Red Ripe Strawberry, and the Big Hungry Bear by Audrey Wood (Children's Play International, 1989)

Rabbit and Hare Divide an Apple by Harriet Ziefert (Puffin Books, 1998)

Sheila Rae's Peppermint Stick by Kevin Henkes (HarperFestival, 2001)

Use with Classroom Activity #2 on page 232.

Strawberry Patterns

Oldies but Goodies

The books listed below currently are not in print but can be found at many local libraries or in used bookstores.

Addition
Moon to Sun by Sheila Samton (Caroline House, 1991)
One Guinea Pig Is Not Enough by Kate Duke (Dutton Children's Books, 1998)
One Plus One Take Away Two by Michael Berenstein (Western Publishing Company, 1991)
On the River by Sheila Samton (Boyds Mills Press, 1991)

Area Code and Telephone Number
The Telephone Book by Maida Silverman (Western Publishing Company, 1985)

Calendar
Sunday Week by Dinah Johnson (Henry Holt and Company, 1999)

Cardinal Numbers 1–10
How Many Feet? How Many Tails? by Marilyn Burns (Scholastic, Inc., 1996)
Numbers All Around Me by Trisha Callella-Jones (Creative Teaching Press, 1998)
Six Crows by Leo Lionni (Scholastic, Inc., 1988)
Six Foolish Fishermen by Benjamin Elkin (Scholastic Library Publishing, 1957)
Six Sticks by Molly Cox (Random House Children's Books, 1999)

Counting, One-on-One Correspondence
How Many? by Debbie MacKinnon (Dial Books for Young Children, 1993)
How Many are in This Old Car? by Colin Hawkins (Penguin Group, 1988)
Ten Dirty Pigs/ Ten Clean Pigs by Carol Roth (North-South Books, 1999)

Counting Backward
Counting Backward by Eileen Christelow (Houghton Mifflin, 1999)
Ten Silly Pigs by Lisa A. Lather (Scholastic, Inc., 1999)

Equal, More, and Less
More or Less a Fish Story by Joanne Wylie (Scholastic Library Publishing, 1984)

Flat Shapes
Shapes by Gwenda Turner (Viking Children's Books, 1991)

Halves
I'll Meet You Halfway by John Schindel (Simon & Schuster Children's Books, 1993)

Left to Right
One Little Monkey by Stephanie Calmenson (Parent Magazine Press, 1982)

Measuring
Measuring by David Kirby (Rigby Education, 1996)
Measuring Penny by Loreen Leedy (Henry Holt and Company, 2000)
Tell Me How Much It Weighs by Shirley Willis (Scholastic Library Publishing, 1999)
You Can Use a Balance by Linda Bullock (Scholastic Library Publishing, 2003)

Oldies but Goodies *(cont.)*

Money
My Rows and Piles of Coins by Tololwa Mollel (Clarion Books, 1999)

Nickel Equals Five Cents by Carey Molter (ABDO Publishing Company, 2003)

Ordinal Numbers
The Stairs by Julie Hofstrand (Front Street, Inc., 1999)

Pairs
Bears in Pairs by Niki Yektai (Simon & Schuster, 1987)

Dirty Feet by Steven Kroll (Parents Magazine Press, 1981)

Left and Right By Joan Oppenheim (Harcourt Childrens Books, 1989)

One, Two, One Pair! by Bruce McMillan (Scholastic, Inc., 1996)

A Pair of Protoceratops by Bernard Most (Harcourt, 1998)

A Pair of Red Sneakers by Lisa Lawston (Scholastic, Inc., 1998)

Socks for Supper by Jack Kent (Gareth Stevens Audio, 1993)

Whose Shoes Are These? by Ron Roy (Houghton Mifflin, 1988)

Whose Tracks Are These? by Jim Nail (Rinehart Publishers, 1994)

Parts of a Whole, What's Missing?
Guess What? by Roger Bester (Value Publishing, 1988)

Just Look by Tana Hoban (Greenwillow Books, 1996)

Look Again by Tana Hoban (Simon & Schuster, 1971)

Tails, Toes, Eyes, Ears, and Nose by Marilie Burton (HarperCollins Children's Books, 1992)

Take Another Look by Tana Hoban (HarperCollins, 1992)

What Is New? What Is Missing? What Is Different? by Patricia Ruben (HarperCollins, 1978)

What's Missing? by Niki Yekiai (Clarion Books, 1987)

Patterns
Pattern Fish by Trudy Harris (Millbrook Press, 2000)

Patterns All Around Me by Trisha Callella-Jones (Creative Teaching Press, 1998)

Same and Different
Bear's Matching Book by Sally Hewitt (Barron's Educational Series, 1995)

Sequencing
Dirty Bird Feet by Rick Winter (Northland Publishing, 2000)

The Turnip by Pierr Morgan (Putnam Publishing, 1996)

What's Next, Billy and Jodie? (Modern Publishing, 1996)

Sorting and Sets
More or Less a Mess by Sheila Keenan (Scholastic, Inc., 1997)

Sorting by David Kirby (Heinemann Library, 1996)

Ten Puppies by Lynn Reiser (Greenwillow Books, 2003)

Time
Winnie the Pooh Tells Time by A.A. Milne (Penguin Group, 2000)

Use with Classroom Activity #5 on page 49.

Number Line

Glue here	Glue here
3	7
2	6
1	5
0	4

Use with Classroom Activity #5 on page 49.

Number Line *(cont.)*

Glue here   Glue here

11   15

10   14

9   13

8   12

Use with Classroom Activity #5 on page 49.

Number Line *(cont.)*

Glue here	Glue here
19	23
18	22
17	21
16	20

Number Line *(cont.)*